Smart Innovation in Agriculture

Smart Innovation in Agriculture

Adele Stewart

STATES
ACADEMIC PRESS
www.statesacademicpress.com

States Academic Press,
109 South 5th Street,
Brooklyn, NY 11249, USA

Visit us on the World Wide Web at:
www.statesacademicpress.com

ISBN: 978-1-63989-729-2

Cataloging-in-publication Data

Smart innovation in agriculture / Adele Stewart.
p. cm.
Includes bibliographical references and index.
ISBN 978-1-63989-729-2
1. Agricultural innovations. 2. Agriculture--Technology transfer.
3. Technological innovations. I. Stewart, Adele.

S494.5.I5 S63 2023
338.16--dc23

Table of Contents

Preface

Agriculture refers to the practice of cultivating soil, developing crops and raising livestock. The broad categories of agricultural products include fuels, foods, raw materials and fibers. Agricultural production can benefit from innovation by increasing productivity, while managing the natural resources efficiently. This promotes sustainability in the long run and reduces the damaging effects of production on the environment. Innovation in agriculture is associated with the management of farms using technologies, such as drones, Internet of things, artificial intelligence and robotics to improve the quality as well as quantity of products, while minimizing the need for human labor in production. Innovations in agriculture technologies encompass developments in precision agriculture, water management technology, farm automation and genetics. These developments enable safer, smarter and more productive farming. The topics included in this book on smart innovation in agriculture are of utmost significance and bound to provide incredible insights to readers. For all those who are interested in this subject, it can prove to be an essential guide.

This book has been the outcome of endless efforts put in by authors and researchers on various issues and topics within the field. The book is a comprehensive collection of significant researches that are addressed in a variety of chapters. It will surely enhance the knowledge of the field among readers across the globe.

It gives us an immense pleasure to thank our researchers and authors for their efforts to submit their piece of writing before the deadlines. Finally in the end, I would like to thank my family and colleagues who have been a great source of inspiration and support.

Adele Stewart

1

Foundation for Successful Innovation

Hugo Campos

1.1 Why Innovation?

> There is nothing more difficult to take in hand, more perilous to conduct, or more uncertain in its success, than to take the lead in the introduction of a new order of things. For the reformer has enemies in all those who profit by the old order, and only lukewarm defenders in all those who would profit by the new order, this lukewarmness arising partly from fear of their adversaries… and partly from the incredulity of mankind, who do not truly believe in anything new until they have had actual experience of it. – Niccolò Machiavelli

The need to innovate is more urgent than ever before. While every economic sector has this need, few are more pressing than agriculture and agrifood systems, as they provide, every single day, food, feedstuff and fibers to humankind. The unforeseen arrival, and far reaching impact, of the COVID-19 pandemic, only reinforces the pressing need to get much better and effective at innovation efforts.

There have been many successful innovations in the field of agriculture to date, leading to increases in farmer productivity and production in many developed and developing countries. As a consequence, the number of people employed as farmers has fallen dramatically in some parts of the world: in 1900, 4 in 10 US jobs were in agriculture, whereas in 2010 this number fell to 2 in 100 (Autor 2014). However, in Africa and Asia, agriculture still remains as a significant source of jobs and income for many people. In Africa alone, despite declining as a share of employment, self-employed farming remains the single largest source of employment, where it

---

This work has been supported by the Bill & Melinda Gates Foundation, investment (OPP1213329), awarded to the International Potato Center (SweetGAINS). It was undertaken as part of, and funded by, the CGIAR Research Program on Roots, Tubers and Bananas (RTB) and supported by CGIAR Trust Fund contributors.

H. Campos (✉)
International Potato Center, Lima, Peru
e-mail: h.campos@cgiar.org

Box 1.1 Main Imperatives Driving the Need to Innovate in Agriculture and Food Systems in Developing Countries

There are more than 820 million people in the world who are hungry, an already daunting figure likely to increase because of COVID-19. Hunger is on the rise in nearly all sub-regions of Africa – nearly 260 undernourished people live in sub-Saharan Africa alone. Asia contains the largest undernourished population – over 500 million people. Nearly 15% of the people in Southern Asia are undernourished. More than 2 billion people, mainly in sub-Saharan Africa and Southern Asia, are micronutrient deficient.

Systems for producing, packaging, and delivering food are responsible for 20–30% of global greenhouse gas emissions, 70% of freshwater withdrawals, and 70% of biodiversity loss.

Twice as much water will be required to produce sufficient food in 2050, but nearly one-third of agricultural production today takes place in water-stressed regions.

Climate change will increasingly stretch current food systems.

accounts for over 50% of all employment in most countries and 35–54% of full-time equivalent-based employment (Jayne et al. 2017). These farmers require to benefit from innovation to boost their chances at success, achieve higher productivities, reduce their workload and gender gaps, and feed more hungry people (Box 1.1). Often, however, new technology and/or ideas that are presented to farmers fail to live up to their initial promise.

There is a critical need to increase the rate of innovation success in agriculture and agrifood systems, to address the so-called *wicked problems* plaguing agriculture associated with climate change and sustainability, as well as to match expectations in terms of adoption and value creation. The goal of innovation in agriculture and agrifood systems is beyond simply achieving financial goals or market share targets. It is also more important even than reaching a target number of users. While these are valid, relevant objectives, they ought to be understood as means to an end: to deliver greater value by means of improved well-being and outlooks, and quality of life for millions of farming families and consumers around the world.

What leads to innovation? Is it simply good technology?

Over 20 years ago, Steve Jobs, former CEO of Apple, said: "One of the things I have always found is that you have got to start with the customer experience and work backward to the technology. You cannot start with the technology and then try to figure out where you are going to try and sell it" (JWWDC 1997). Technology is simply a driver of successful innovation. It operates within an ecosystem of individuals and nonprofit, government, and private organizations focused on the social and economic use of new products, processes, and organization forms. Technology alone will not produce enough nor sustained success in any sector, including agriculture. I would posit that, increasingly in today's and tomorrow's world, technology alone does not represent the main determinant of the perceived value innovations create for its users.

Is the answer a big budget?. Do more money and larger teams mean more innovation?

Large budgets are not necessarily a good predictor of innovation success. A myriad of well-funded innovation efforts has fallen short of expectations or outright failed in the market. Case in point: Motorola's failed launch in the 1990s of Iridium–a mobile telephone service providing global coverage that cost investors over $5 billion. While Iridium was expected to capture millions of customers, by the time it filed for bankruptcy protection, it had only acquired about 55,000. It was finally purchased for the discounted price of $25 million. Many modest budgets, in contrast, have managed to deliver high value to both investors and users. See Box 1.2 for a cassava example.

If it's not technology or money, what leads to successful innovation?

One pivotal part of the answer is *people*. People are the most effective predictor of success of any innovative undertaking. People cannot thrive, however, in organizations that stifle their creativity and experimentation. Organizations must support, encourage, and foster a culture that enables employee innovation, or they risk crippling the talent they hired to produce results in the first place, leading employees to look for a brighter outlook and professional development elsewhere. To succeed, organizations must invest in building a system that both nurtures human innovation and takes the user into account.

Most innovation failures stem from failing to recognize that innovation is both an economic and social process rather than simply a technical one. Many innovation endeavors fail despite funding availability, talented people, strong technology, and sincere intentions. Ultimately, innovation success depends on whether the value developers *think* they are delivering matches the *actual* value users[1] find in the product or service. The business model associated with any innovation is also pivotal to fulfilling users' expectations and determining if an organization survives.

Why do we have to innovate? Isn't maintaining business as usual good enough?

The simple, straight, and honest answer is *no*. Failure to address disruptive change and implement business model innovations has led to the demise of many successful companies and organizations, in both developed and developing countries.[2]

[1] For the sake of simplicity, throughout this chapter, the term "users" will also encompass other parties toward whom innovation efforts are targeted, such as customers, clients, adopters, end-users, and beneficiaries.

[2] The longevity of corporations is rapidly declining in the United States. In order to survive, they must innovate more rapidly than ever before. The 33-year average tenure of companies on the S&P 500 index in 1964 shrank to just 24 years by 2016 and is forecast to drop to just 12 years by 2027. Even large companies such as Alcoa, DuPont, and Yahoo all left the S&P 500 index in the period 2013–2017. Companies like these have been replaced by firms like Facebook, Under Armor, and PayPal. While the chance of remaining in the S&P 500 index for the first 5 years after being listed before 1970 was over 90%, companies listed from 2000 to 2009 only had a 60% chance of maintaining their grip on that rank (Anthony et al. 2018). A similar trend pervades all markets, including agriculture, both in developed and in emerging economics.

The pace of change is increasing rapidly. While this shift can be cause for concern in traditional organizations, it also presents an opportunity – a bright outlook for those bold enough to launch innovations that contest the *status quo*.

Innovation is the key to how an organization not just survives but thrives instead.

Box 1.2 Innovative Ways to Create Value Out of Agricultural Byproducts

Cassava is the main crop in Africa on a wet weight basis, with Nigeria being the world's largest cassava-producing country[3]. Cassava peels make up 20% of the whole root, but are discarded during processing. The peels amount to nearly 40 million tons per year in Africa alone, giving cassava a bad name as an environmental polluter with the mountains of waste around processing centers. To create a business opportunity out of this undesirable byproduct, the International Institute of Tropical Agriculture has developed high-quality cassava peel (HQCP) feed ingredients from wet peels. This innovation enables rapid water removal and accelerates the elimination of hydrocyanide. The intermediate product (60% dry matter, up from 30% in fresh peels) is safe for livestock to consume and stable for up to a week and can be sun-dried or heat-toasted to a storable product (90% dry matter). This can be done any time of the year in a small- and medium-scale setup or flash-dried in a more industrial case (see http://bit.ly/2j7bRu3). Since three tons of fresh peels yield about one ton of HQCP, Africa's cassava peel waste could generate at least 12 million tons of HQCP annually – equivalent in metabolizable energy (ME) to 8 million tons of maize thus spared for direct human consumption. In addition, there is willingness to pay for HQCP; for example, when maize prices reached \$300, HQCP was being purchased for \$150. This ratio holds for wide price bands.

The huge value creation of this high-impact innovation provides an alternative source of feedstuff, protects the environment, and provides new income sources to smallholders producing cassava. It has been supported by the CGIAR Research Program (CRP) on Root Tubers and Bananas (RTB), and it leverages the expertise of several private and public partners in Nigeria, such as the National Office for Technology Acquisition and Promotion (NOTAP), Raw Material Research and Development Council (RMRDC), Bank of Industry, and SingleSpark® from the Netherlands, makers of FeedCalculator®.

Source: Dr. Iheanacho Okike, International Institute of Tropical Agriculture, (i.okike@cgiar.org)

[3] http://www.fao.org/faostat/en/#home [verified on October 11th, 2019].

1.2 Toward a Working Definition of Innovation

Since more than a century ago, the seeds of these ideas were already present. In Austrian economist Joseph Alois Schumpeter's book, *The Theory of Economic Development* (1911), the concept of "creative destruction" was articulated as a process "that incessantly revolutionizes the economic structure from within, incessantly destroying the old one, incessantly creating a new one." He also described innovation as the cause of market dislocations, which enabled new firms to displace old firms from markets (Schumpeter 1942).

Schumpeter pioneered the idea that competition should not be based only on price, but instead on capabilities and performance. He was the first known economist to shift the basis of competition from the ability to reduce costs to the capability to innovate.

Building on these ideas, for the purposes of this book, we will use Scott Berkun's definition of innovation. Berkun, author of *The Myths of Innovation* book, states that "innovation is significant, positive change." Innovation is something you work toward. It should be the end result or outcome. It is the sum of an organization's efforts to keep its value proposition to customers and end-users both relevant and compelling.

Though such definition of innovation appears to be unexpectedly unassuming and simple, be mindful that the articulation of all the components leading to the successful practice of innovation, and the ability to claim success in terms of significant value actually delivered to customers, is a remarkably complex, yet attainable, endeavor.

1.3 The Deep Relationship Between Innovation, Design, and Human Needs

The growing awareness that technology is only one driver of successful innovation has led to the development of a more human-centered approach to innovation. This approach is strongly anchored in a deep understanding of customers and other stakeholder needs, preferences, non-spoken and poorly articulated perspectives, and emotions. Here the attention is not on focus groups or on collecting and analyzing a deluge of marketing data. Instead, human-centered innovation identifies the sources of *meaning* for users.

Because users have a rather limited ability to articulate their unmet needs, innovators increasingly borrow the perspectives of designers to capture such needs. This goes well beyond purely aesthetic aspects and instead builds on empathy, visualization, enlightened failure, and market experimentation. A design perspective aims to understand the behavior that people struggle with while living their lives, bringing a human-centered perspective to innovation efforts. Only through the willingness to immerse yourself in the lives of users can you uncover their constraints and frustrations with non-existent or limited solutions. That immersion simply cannot happen in your office or within the boundaries of your organization. Instead, you must empathize with the pain, disappointment, and frustration your users endure in their daily lives.

In the agricultural and agrifood sector, there have been diverse attempts to understand what the needs of farmers, members of supply chains, and users look like and to involve them in the technology design and innovation processes. However, a deliberate and explicit consideration of human-centered design to address how technology fits into the social, cultural, and emotional context of its users is not common practice. This is unfortunate, as such considerations would only increase the likelihood of success of innovation efforts, as well as the ability of innovation practitioners to truly apprehend how value is perceived by farmers and other participants in food systems. Fortunately, several diverse human-centered methods to innovate have been already developed in other domains. The following section briefly describes one approach – design thinking.

1.3.1 A Primer on Design Thinking

One highly effective approach to create human-centered products and services is design thinking.[4] It is typically structured as three overlapping phases:

- Inspiration
- Ideation
- Implementation

Typically, these phases proceed sequentially, although sometimes design thinking loops back onto itself as solutions are tried out. It is quite different from highly linear or more rigid approaches to innovation.

Design thinking shapes the experience of both innovators and users. It recognizes organizations as human groups and factors in the importance of emotions. It encourages learning and engagement. By closely involving users in the identification of the opportunity and its solution, design thinking garners a wider commitment to change (Liedtka 2018). Among its stand-out main characteristics, design thinking is human-centered, possibility-driven, option-focused, and iterative. Design thinking starts with real human beings instead of demographic data, and before even attempting to generate solutions, it purposely delves into the lives, challenges, and circumstances of the people whose lives we aim to improve. It is possibility-driven since it addresses the question, "What if anything were possible?" and it focuses on developing multiple options before zooming in on the most appropriate one. Furthermore, design thinking is iterative, as it sails through several rounds of prototypes and multiple rounds of real-life experiments to refine ideas and potential solutions (Liedtka et al. 2017).

Design thinking is a highly effective, pragmatic innovation and problem-solving approach.

---

[4] A detailed discussion about design thinking goes beyond the scope of this chapter. Books such as *Change by Design* by Tim Cook and *Creative Confidence: Unleashing the Creative Potential Within Us All* by Tom Kelley and David Kelley provide a more extended perspective on design thinking.

1.3.1.1 Inspiration

During the inspiration phase, multidisciplinary teams look for user satisfaction gaps. These gaps are sometimes referred to as the "the pain of the customer." They are different, however, from the "voice of the customer." The most effective way to apprehend satisfaction gaps is to learn directly from the people you seek to help. Identifying these gaps cannot be achieved through surveys, interviews or focus groups, or demographic data, though. Instead, you must immerse yourself in the lives of users to deeply understand their needs, pains, and aspirations.

One option to achieve that, among many alternative ones, is to develop a detailed map of the journey users take in their quest to address their satisfaction gaps. An ethnographic analysis of all the contact points occurring between users and goods/services is also an effective approach. The construction of empathy maps and trajectory maps are also effective approaches during this stage. The focus of inspiration should not be on data collection and analysis, but rather on gaining insight about what makes for a meaningful customer journey (Liedtka 2018).

1.3.1.2 Ideation

Ideation ensues inspiration, and it aims to detect unarticulated gaps between what users require and what they actually receive. This phase generates, develops, and tests entry points for design. Field observations and research are distilled into insights to discover solutions or drive change.

The inspiration and ideation phases are about divergent thinking, where on-purpose insights are made to clash against each other. Openness, empathy for others, sheer curiosity, and the ability to learn by doing are essential to this process. At this stage, the visual representation of concepts through early draft prototypes can help explain complex thoughts and help the team gain insight. During ideation, people identify themes, create insight statements, and attend co-creation workshops where users and innovation developers jointly work together to establish solutions.

1.3.1.3 Implementation

Implementation is the ensuing phase, when the innovation solution actually gets developed. The best concepts from the ideation stage should point to a clear way forward through prototyping. Prototyping uncovers potential implementation challenges and unintended consequences of products or services.

Prototypes anticipate how people understand the potential solutions to their satisfaction gaps. The team develops several iterations of concept prototypes. Later on, functional prototypes are tested, and a "looks like" design model is agreed upon. Prototyping plays an especially important role in developing countries where innovation ecosystems are generally weak or incomplete. Prototyping enables innovators to experiment, learn, and adapt concepts, so they can quickly and cheaply refine the product into something even better.

Successful design thinking does not just end with a product or service that simply reduces the pain of the user. The solution must then be communicated to a wider audience through powerful, compelling narratives and storytelling for wide product or service adoption to take place.

A word of warning – design thinking can create substantial tension in organizations and companies accustomed to linear, milestone-based product development and innovation. One of the tenets of design thinking and human-centered design is the need to achieve a close rapport with users – something at odds with a more structured view of developing new products and services.

Additionally, design thinking not only embraces but actively seeks failure as a key learning approach. Yet, many organizations have a near zero tolerance for failure. That means that the implementation of a human-centered approach in organizations requires strong leadership and full organizational support, or else it is likely to fail, once again leading to innovation efforts falling short of expectations (Bason and Austin 2019).

1.4 Innovation Is Different from Research and Development

Often innovation and research and development are referred to as if they are interchangeable concepts. Nevertheless, their meaning, scope, and purpose are quite different: research and development (R&D) comprise creative and systematic work undertaken in order to increase knowledge – including knowledge of humankind, culture, and society – and to devise new applications of already available knowledge (OECD 2015). This differs in many ways from the working definition of innovation provided in Sect. 1.2.

The difference between the two can create misunderstandings, as well as conceptual, practical, and very expensive mistakes. Though in theory, corporate R&D spending relates to increased innovation and the growth of revenues and profits, that's not always the case. The magnitude of R&D expenditures is only moderately related to the number of innovative products/services launched or even those that gain high market share and profits.

To some extent, confusing R&D and innovation arises from a narrow, technology-driven stance on innovation. Innovation, by default, encompasses a much wider scope than R&D, and as Gina O'Connor from Babson College aptly puts it, innovation involves three diverse capabilities: discovery, incubation, and acceleration. In many cases, the resources, investments, and time required during the incubation and acceleration stages far exceed those needed to develop technologies in the first place.

Indeed, within profit-seeking firms, R&D and market success are two different things. *Strategy &* – a business unit within PricewaterhouseCoopers – was unable to find a statistically significant relationship between R&D spending and sustained financial success when it analyzed the top 1,000 most innovative companies over 12

Table 1.1 Top R&D spenders and innovative companies in 2018

Most innovative companies		Top R&D spenders	
Rank	Company	Rank	R&D spending (US $ billion)
1	Apple	1	Amazon (22.6)
2	Amazon	2	Alphabet (16.2)
3	Alphabet	3	Volkswagen (15.8)
4	Microsoft	4	Samsung (15.3)
5	Tesla	5	Intel (13.1)
6	Samsung	6	Microsoft (12.3)
7	Facebook	7	Apple (11.6)
8	General Electric	8	Roche (10.8)
9	Intel	9	Johnson & Johnson (10.6)
10	Netflix	10	Merck (10.2)

years. Spending on R&D was also found to be unrelated to growth in sales or profits. Moreover, the top 10 most innovative companies are rarely the top 10 spenders on R&D (Viki 2016).

Which main aspects separate the most innovative companies from less innovative ones?

More innovative companies show:

- Close alignment between innovation and business strategy
- Company-wide cultural support for innovation
- Close involvement by leadership with innovation programs
- Deep understanding of insights from end-users

This trend was again observed in 2018 (Table 1.1), when only two out of the five most innovative companies, namely, Amazon and Alphabet, also ranked among the five top R&D spenders.

Analyzing R&D intensity – a metric defined as the ratio between R&D expenditure and total revenue – sheds light on the effectiveness of innovation efforts. Dramatic differences can be verified even among the most innovative companies. Although Amazon, Alphabet, Microsoft, and Tesla show a large R&D intensity, within a 12–15% range, in the case of the most innovative company of 2018 – Apple – its R&D intensity only reached 5%. Furthermore, in the case of Nokia, despite a high R&D intensity in 2018 (21%), it ranked lower than 20th in terms of innovation.[5]

It might be a little unfair to compare firms across diverse industries and R&D intensity landscapes. Regardless, such comparison sheds light on the actual nature of the relationship existing between R&D efforts and the ability to translate them into products and/or services that are actually adopted by users and therefore create value.

Markovitch et al. (2015) provide reinforcing evidence regarding the lack of a direct relationship between R&D efforts and innovation outcomes: Using data from 141 US firms across a decade of data, they could not find any statistically significant relationship between a firm's investments in basic, exploratory R&D (measured by each firm's number of patents over the previous decade, weighted by how scientifically novel they were) and the firm's stock market value.

This should not be a surprise once the meaning and purpose of innovation are understood. R&D can only contribute to the development of goods or services that are technically superior to previous alternatives.

What about other examples?

As you will see throughout this book, successful innovation requires much more than technical advantage. If this were the sole factor driving success, the Betamax system developed by Sony would have become the dominant design and the commercial winner in the video cassette recording market in the 1980s. Instead, it rather quickly became obsolete.

A successful example of limited R&D expenses related to a successful innovative product is the Japanese company Nintendo, which developed the Wii home video game console from off-the-shelf components. It achieved remarkable commercial success with a technologically inferior product compared to those of competitors. And it did it in a market where Sony and Microsoft spent much more on cutting-edge gaming consoles.

Yet another well-documented example about the non-existent or weak relationship between R&D spending and innovation success is the garment industry in the Philippines. Despite no access to formal R&D or academic collaborations, it has introduced incremental innovations in both products and processes, through reverse engineering and combining knowledge on new ways to successfully innovate and remain competitive in global markets (Rosellon and Del Prado 2017).

This is not meant to discourage R&D efforts. Many successful innovations have required long and expensive public or private research efforts. Many companies have translated large R&D budgets into successful innovations that are rapidly adopted. The takeaway message here is that an expensive R&D effort does *not guarantee* innovation success. The opposite also holds true: it is perfectly feasible to achieve large returns from innovation and become a successful organization in the field of innovation with a rather limited R&D investment. Furthermore, Open Innovation (see Chap. 3) is a very effective approach to increase the ratio of success of innovation efforts.

Innovation can succeed regardless of an organization's size, the nature of its users, or the economic sector within which it operates. That's encouraging news for young start-ups, small, and cash-strapped organizations. If they have the talent and boldness to lead, design, and execute innovation plans, they have a high likelihood of successfully challenging and even defeating older, larger, and much better-funded organizations.

1.5 Does Innovation Only Pertain to Profit-Seeking Organizations?

Most of the insight, experience, and literature relevant to innovation relates to for-profit efforts where companies and organizations innovate to remain in business. They grow and adapt to the needs and desires of customers and markets along the way. However, that does not mean that the natural domain of innovation is only within profit-seeking organizations. Leewis and Aarts provide an updated perspective of innovation adoption in agriculture in Chap. 4 of this book.

Nonprofit organizations, regardless of their role, size or geographic footprint, are also under increasing pressure to innovate. Their users and beneficiaries have needs that continuously evolve. They must adapt to the needs and expectations of their donors and funding agencies, as well as to the increasing scrutiny from governments, international bodies, and philanthropic organizations.

From this perspective, this chapter argues *that nonprofit organizations already run a business*. They might not seek profits but, nonetheless, they remain a business (Box 1.3). Regretfully, even today, the word "business" raises concerns, and it is not taken well within some university and other knowledge-seeking organizations.

Are there good examples of an innovative nonprofit?

Several Boxes provided throughout this book describe successful nonprofit innovations. A further example of an innovative nonprofit model is the Innovation Lab set up by Oxfam.[6] In a systematic way, it addresses core development challenges that have potential for market systems solutions. Market-based ideas are analyzed, and pilot projects are run to make the solutions better.

BlocRice is an innovative program launched by Oxfam in Cambodia aimed at increasing rice farmers' leverage to secure fairer prices. It is based on digital contracts established among rice agricultural cooperatives, exporters in Cambodia, and

Box 1.3 The Actual Meaning of the Word Business

Often the word "business" encounters skepticism, diverse opposition, and even a degree of disdain in academic and nonprofit organizations. This is unfortunate, as the actual meaning of this word has very little to do with corporate greed or utter disregard for anything other than a ruthless pursuit of profits.

The word "business" is derived from the old English word *bisignes*, "attend to, be concerned with, be diligent." Its actual meaning is "moving towards a goal, to be determined about serious action." We advocate the use of this word because it conveys a positive activity worth pursuing.

buyers in the Netherlands. Yet another very good example of innovation in a nonprofit organization is Mercy Corp's Social Venture Teams, which acts as an internal incubation and acceleration lab. Its purpose is to develop innovations able to be scalable into business in emerging markets. Chapter 6 by van der Velde et al. in this volume delves into several, additional successful examples of innovation within the nonprofit domain in developing countries.

The increasing relevance of innovation for nonprofit organizations is best reflected by emerging rankings that highlight the most innovative nonprofit organizations.[7] For instance, the magazine *Fast Company* has added nonprofit organizations to the large number of sectors it ranks on innovation.

Does this extend to the public sector?

The need to innovate also reaches the public sector since government delivers, as a monopoly, services to millions of citizens who sustain it with taxes. The public sector represents the main economic force in many countries, particularly in emerging ones. As global standards of transparency, accountability, and value for money become more stringent, governments are also increasingly pressured to innovate.

Public sector innovation in developing countries is even more pressing and relevant, since public sector expenditures represent a larger proportion of gross domestic products (GDP) in many such countries than in developed ones. Furthermore, innovation in the agriculture and agrifood domains is particularly critical in developing countries, as their economic footprint is much larger than in developed ones. For instance, agriculture, food, and related industries accounted for about 37% and 5% of total GDP in Ethiopia and the United States in 2017, respectively. In other words, properly executed innovation in agriculture and agrifood systems has much greater leverage, spillover power, and a larger ability to create additional value and new jobs in developing countries than in developed ones.

While many governments are forced to adapt to a maelstrom of change, progress is often ad hoc rather than reliable, reactive vs. deliberate, and sporadic instead of systemic. The public sector has not yet ensured that innovation leads to consistent and reliable options that can deliver better outcomes and greater impact.

OPSI,[8] the OECD Observatory for Public Sector Innovation, provides tools, resources, and models for public sector innovators. These are important because public-service institutions need to learn to innovate and to pay more attention to social, technological, economic, and demographic shifts. It's the only way to cater to the increasing needs of their progressively empowered citizens.

What's a good example of innovation in the public sector?

There are some encouraging examples from the public sector, though. One such example is NESTA (https://www.nesta.org.uk/) in the United Kingdom,[9] which represents a remarkably successful example of developing innovation in the public

[6] NESTA was formerly the National Endowment for Science, Technology and the Arts, which evolved into an independent organization in 2012.

domain. In addition to developing creative economy and innovation policies, NESTA impact includes, but is not limited to, health, education, and the arts.

In today's world, the need to innovate spans the entire continuum from private to public sector – profit-seeking and nonprofit organizations. At its core, innovation is a way of life, an attitude based on continuously challenging the *status quo*, rather than merely reacting to changes in customers, regulations, competitors, and markets.

1.6 Disruptive Innovations

Perhaps one of the greatest misunderstandings about innovation comes from the concept of "disruptive innovation." When it works, it can allow smaller, less-funded organizations to disrupt the market share of established, incumbent ones (Christensen et al. 2015). Although there are certainly large organizations able to drive smaller organizations out of business, when disruption occurs, it can allow a proverbial David to overtake Goliath.

Products or services in and of themselves do not represent disruptive innovations, however. The concept refers to the *trajectory* of a product or service along the value curve rather than about a discrete product or service. Disruptive innovations need not be breakthrough innovations. Instead, they consist of products and services that are simple, accessible, and affordable (Dillon 2020).

1.6.1 How Do Disruptive Innovations Unfold?

Incumbent organizations usually focus on improving the value delivered to their most demanding customers. This follows the logic that these customers are the most profitable. The result is that the needs of only a fraction of users are met. Other customers are left inadequately served (incumbent's sustaining trajectory in Fig. 1.1). These overlooked and/or non-consuming customers represent a significant opportunity to provide more than expected value at a (relatively) low price.

Since incumbents tend to focus on the most profitable, high end of the market, they usually fail to defend the least profitable segment as vigorously. This allows new entrants – the disrupters – to move up-market, improving value and quality while keeping the features that enabled them to challenge the incumbents (Entrant's disruptive trajectory in Fig. 1.1). Innovations tapping into non-consumption are particularly relevant in developing countries, since they represent a means to alleviate poverty. They can open new markets, becoming the main building block toward greater prosperity, as argued by Christensen et al. (2017).

Disruptive innovations originate either from the low end of markets or from tapping into products and services not yet consumed (Wunker and Farber 2019) (Box 1.4). One private sector example is low-cost air travel, a hugely successful disruptive innovation built upon addressing non-consumption (Taneja 2017). Many low-cost air carriers, such as Southwest in the United States, EasyJet and Ryanair in

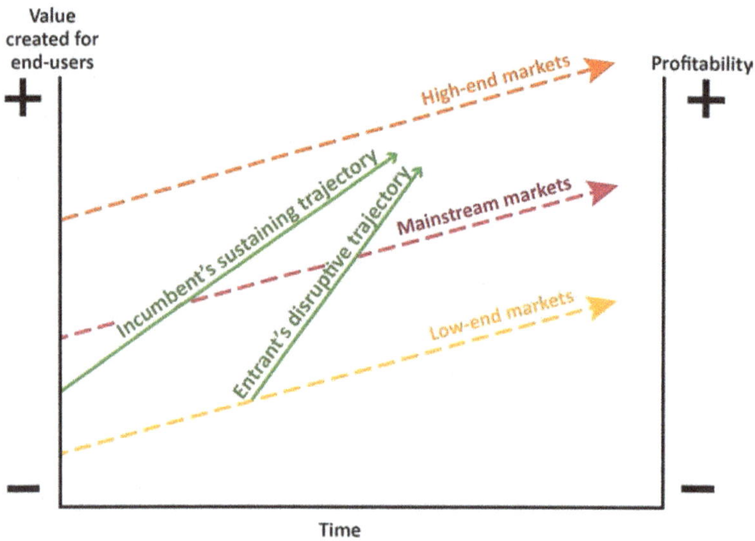

Fig. 1.1 Trajectory in the market of disruptive innovations. (Modified from Christensen et al. 2015)

Europe, IndiGo in India, Azul in Brazil, Lion Air in Indonesia, and Sky Airline in Chile, have successfully challenged incumbent, larger airlines.

These no-frill air carrier fleets usually have only a single type of aircraft to reduce maintenance costs. They fly to smaller, secondary airports which are cheaper to operate.[10] The cheap fleets reduce staff by issuing tickets and boarding passes digitally. They charge high fees for checking in with a staff person. This shift has enabled these companies to tap into a mass of consumers who would otherwise use cars, trains, or buses or just not travel at all. In other words, low-cost carriers do not necessarily compete against other airlines. Instead, they cater to the very large number of customers that would have not previously considered travelling by air.

In the case of low-cost air travel, the financial advantage of innovation came as a result of improving the quality of life of its users by making air travel affordable.

A relentless focus on the needs of users should also be aimed at by agricultural and agrifood systems innovations. If individual innovators address the unmet needs of users, returns on investment will increase.

In Africa, innovators who address the daily unmet needs of low-income consumers have a larger likelihood of success than if they chased after higher margin opportunities arising from the growth of the middle class. This market creation mindset is behind the shift from poverty to prosperity in Taiwan, South Korea, Singapore, and Hong Kong.

---

[7] In the case of London, Luton airport offers much cheaper fees to air carriers than flagship airports such as Gatwick and Heathrow. In the state of California, Oakland airport offers cheaper landing and taxiing fees than its neighbor, San Francisco airport.

Box 1.4 Tapping into New Markets: Rapid Diagnostic Tests for Malaria
Malaria is a severe public health issue. According to the World Health Organization (WHO), in 2016 there were over 215 million cases of malaria globally, which claimed nearly 450,000 deaths each year, out of which nearly 200 million cases and 400,000 casualties occurred in Africa. It remains as the single most deadly global health disease.

Blood tests are unaffordable for many Africans. Others are reluctant to go to hospital and be tested for malaria using traditional diagnostics. Instead, they rely on self-medication to treat potential cases. This creates yet another public health burden because many of the misdiagnosed cases are treated with antibiotics. This unneeded treatment accelerates the building up of antibiotic resistance, which has already rendered unsuccessful, previously highly effective drugs such as chloroquine and artemisinin.

The Baltimore, USA-based startup company Fyodor Biotechnologies saw an opportunity in this problem and developed a low-cost urine test for malaria. For less than $2 and without the need for access to health workers, this test provides an accurate diagnosis in less than 30 minutes. It is simpler to use than testing blood samples. It has already been launched in Kenya, and plans are in place for its commercialization in other markets.

Another reason to focus efforts at the base of the pyramid, as discussed elsewhere in this volume by van der Velde and coauthors in Chap. 6, is that the actual growth of the African middle class has been much less robust than forecast (Christensen et al. 2017).

An example that illustrates how this principle works in action is the development of sweetpotato cultivars enriched in vitamin A, known as Orange-fleshed sweetpotato (OFSP) by the International Potato Center, which explicitly examined the unmet needs of end-users, many of whom were living in poverty. Such cultivars represent a rich, affordable source of beta-carotene, which the body converts into vitamin A (Low et al. 2017). Scientists at the International Potato Center challenged the conventional wisdom concerning food-based approaches and institutional barriers, to address the unmet need of having a diet with proper levels of vitamin A, particularly in children and lactating mothers. To date, a multi-partner, multi-donor initiative based on OFSP has already reached over 6 million households in sub-Saharan Africa, most of them representing the base of the pyramid.

Table 1.2 Some traits of the exploitative and explorative domains observed in ambidextrous organizations

	Ambidextrous organizations	
	Exploitative business	Explorative business
Type of innovation	*Sustaining innovation*	*Disruptive innovation*
Strategic intent	Cost, profit	Learning, growth
Critical tasks	Operations, efficiency	Nimbleness, new business models
Culture	Efficiency, risk-averse, quality	Speed, risk-taking, flexibility
Leadership style	Authoritative, top-down	Visionary, involved

Adapted from Frederik (2015)

1.7 The Dilemma of Exploration Versus Exploitation

One of the most intellectual and practical challenges that organizations face is how to balance the goods and services they currently deliver to customers, against the need to develop the future innovations their continued existence depends upon.

Many once-dominant companies have faltered when developing products needed to address market changes – to either retain current customers or gain new ones. Kodak excelled at analog photography and was the uncontested global leader for decades on end. It failed, however, to make the leap to digital cameras and adjust its business model to succeed in an increasingly digital world.

Organizations must create a *dual capacity* to exploit their current business while exploring new opportunities. These "ambidextrous organizations" have been extensively analyzed by Charles O'Reilly of Stanford University and Michael Tushman of Harvard University. Ambidextrous organizations pursue sustaining innovations that increase business success, instead of solely focusing on creating new markets. Typically, they have most of the traits shown in Table 1.2 and are able to pursue disruptive, exploratory innovations which are critical for the organization's future (O'Reilly and Tushman 2004).

Much of the discussion on how to implement ambidexterity examines structural aspects, such as the creation of separate units to pursue new opportunities while keeping the same general manager to lead both the new unit and the parent company (O'Reilly and Tushman 2004). However, there are equally if not more relevant aspects to consider, such as having strong, shared values within the organization, which enables corporate headquarters to relinquish some autonomy to divisions without losing overall alignment and strategic fit.

Govindarajan (2016) recently developed a practical framework to allocate time and resources to the competing and opposing demands of managing today's needs and tomorrow's possibilities:

- *Manage the present core business at peak efficiency and profitability.*
- *Identify and abandon old and irrelevant practices, ideas, and attitudes.*
- *Convert breakthrough ideas into new products/services and business.*

Becoming proficient at both current business and future exploration is no small feat, though. It has only been achieved by a few companies. Recent work by Haan et al. suggests that no more than 2% of companies achieve it.[11] Among the features shared by such a select group of companies, the following stand out: excellence at both exploration and exploitation; retaining an "outside-in" focus even when successful; embracing necessary disruptions – even if painful; and having in place a clear model for renewal.

1.8 Innovation and Failure[12]

Most innovations launched to the market by companies fall short of revenue expectations. Approximately 80–85% of all fast-moving-consumer-good launches fail, according to 2018 research[13] by Nielsen, the global measurement and data analytics firm. Though there is no equivalent data known on nonprofit organization innovations, it is unlikely that the opposite would apply to their goods and services.

This points to substantial shortcomings in the ability to match value propositions with the actual needs of customers and users. While the technology itself can certainly go wrong in the design process, the difficulty of designing the right business models – vs. simply the right technology – remains one of the most challenging components of innovation success.

How could it be that the vast majority of commercial and social innovations fail – falling short of their profit, adoption, and/or impact goals?
How can we better articulate the social, emotional, and economic gains brought about by innovations so that their adoption rates improve?
Why are people so reluctant to embrace change?
Why do most organizations, both private and public, struggle in their innovation efforts?

These are critical questions that every individual and organization genuinely interested in embracing innovation need to consider.

In the agricultural realm, the story of coffee illustrates how innovation can lead to tremendous success despite failure. Today, coffee[14] is a regular part of the lives of

---

[8] Throughout this chapter, we will refer to failure as the inability to reach expected sales, profit targets, adoption, or market penetration of innovations launched to the market. Though it is a rather narrow definition, it is a pragmatic, results-driven one and consistent with the book #x2019;s views on innovation.

[9] The late Calestous Juma, formerly of Harvard University, discusses the history of coffee in his book *Innovation and its Enemies* (2017), shedding light on the tensions and complex interactions arising among innovations, preexisting products, cultural landscapes, and social fabrics.

people in most of the world. For many of us, it represents nothing less than a "must have" to start a productive day.

Yet, hundreds of years ago, coffee was banned, several times, in many parts of the world. In Mecca, it was banned because of the fear that coffeehouses could become sources of social agitation against religious authority. In Constantinople, during the reign of Sultan Suleyman, a fatwa against drinking coffee was issued in 1543. In Europe, the situation was not any different as coffee and/or coffeehouses were once banned or curtailed by King Charles II in England, King Frederick the Great in Prussia, and the Swedish Parliament!

Obviously, tremendous social and cultural changes have shaped our perceptions of coffee. Only two decades ago, coffee shops in the United States were generally small, locally owned places where people gathered over poor-quality, cheap coffee. Along came Starbucks, which changed the game by introducing Americans to high-end, European-style coffee drinks that are now a regular part of the lives for millions of people every day. How was Starbucks able to innovate so successfully in a relatively short period of time? A main driver has been its ability to adroitly cater to the needs of its customers and to the job they need to get done while drinking coffee. For instance, Starbucks does not just provide coffee for its customers. Instead, it provides a "third place" outside their homes and workplaces, which offers not only coffee but also a secure, warm, and welcoming environment where people can gather and connect.

As of 2018, Starbucks runs over 28,000 coffee shops and employs nearly 300,000 people worldwide. To put its global team size in context, that's more people than the population of about 150 individual countries.[15] Yet, Starbucks is having to innovate again. Starbucks Reserve is the coffee giant's newest initiative of specialty outlets which are meant to compete with third-wave coffee shops that source unusual or small-lot beans and train baristas in hand-poured preparation methods. Ever-evolving customer needs have to be taken into account, or Starbucks may soon find itself going the way of Eastman Kodak.

The learnings from the development and adoption of a globally successful innovation such as Starbucks in the coffee sector can be widely applied to a wide range of sectors, regardless of the technology they use, the sector they work in, or the geography they serve.

The launch of new technologies represents social experiments. There are many psychological, social, and economic factors that individual and organizational innovators must understand. The sheer success and global adoption of Starbucks illustrate how success can be achieved if an organization takes into account the emotional and social underpinnings of its end-users.

The takeaway here is that tensions around innovation arise not so much from either the potential benefits or pitfalls an innovation represents. Rather, tension comes from hidden feelings and sentiments of insecurity, since innovations challenge seemingly stable economic interests and social institutions.

One positive way to frame innovation failure is: "failure is the experience that precedes success." There is rarely successful innovation without a fair share of failure along the way. Every successful innovator deals with the frustration, pain, and anxiety associated with numerous setbacks. I actually invite you to embrace the concept of *clever failure*, namely to fail often, cheaply, and as early as possible in the innovation process. Because failure is inherent to innovation, organizations should embrace it and help employees brace for it, and provide mechanisms to de-risk it instead of shying away from it. Take pains to understand why consumers rarely adopt or purchase innovations at the rate developers expect. A basic understanding of behavioral economics provides insight and can help organizations plan for the way human beings act, adapt, and react.

1.8.1 Loss Aversion

Loss aversion refers to when losses loom larger than gains.[16] This asymmetry between the power of positive and negative expectations or experiences has an evolutionary history. Species that treat threats as more urgent than opportunities have a better chance to survive and reproduce (Kahneman and Tversky 1979).

What has loss aversion to do with innovation failure? A lot. At its core, loss aversion is an expression of fear. Losses elicit stronger, more visceral feelings than comparable gains. When our customers, used to the value that a good or service provides, are faced with innovative alternatives, they might rather stick with the one they feel comfortable with to reduce the risk of losing the value they are used to receive.

This has obvious implications for the rate at which innovations are adopted and purchased. Agriculture provides excellent examples of how loss aversion can significantly obstruct the adoption of technologies. Ward and Singh (2014) elegantly demonstrated that more loss-averse farmers are less likely to switch to new rice cultivars, although the new cultivars clearly outperform older, popular varieties under both normal and drought conditions. No wonder that many new cultivars, across many different crops, fall short of expected adoption, and therefore do not achieve profit or impact goals.

1.8.2 **Status Quo** *Bias*

Status quo bias refers to the tendency to stick to the current status of affairs. One implication of loss aversion is that individuals have a strong tendency to remain at the current *status quo*. This has been extensively demonstrated through decision-making experiments (Kahneman et al. 1991).

Status quo bias explains why we are biased toward the default option: Most people tend to use their default Internet browser instead of installing a new one which

---

[10] Richard Kahneman, 2002 Nobel Memorial Prize Winner in Economic Sciences.

may provide more speed, improved security, and a better navigating experience. *Status quo* bias can make people stick to their usual choices even when not in their best interests. Recently, Karl et al. (2019) demonstrated that participants in health insurance programs in the US that provide different financial benefits could be classified from very low to very high *status quo* bias categories. A high *status quo* bias was associated with a higher rate of physical inactivity, a higher sum of unhealthy lifestyle factors, and a higher BMI.[17]

What does it take to get end-users to adopt a new innovation?

Innovative goods or services often require consumers to change their behavior. Organizations trying to innovate frequently fail to fully appreciate just how daunting that task is that they face. Consumers assess innovations in terms of what they gain and lose relative to existing products. John Gourville of Harvard University refers to this as the "curse of innovation." Generally, consumers overvalue the existing benefits of an entrenched product, whereas innovators overvalue the benefits of their new products/services. The result is a large mismatch between what developers think users desire and what users really want. This widening gap between the two groups increases the likelihood of failure for even the most innovative new products.

Gourville (2006) suggests avoiding this mismatch by avoiding the launch of innovative goods or services that demand substantial behavioral change. He argues that organizations should target non-consumption and strive for benefits that outrun losses at least by a factor of 10 as a rule of thumb. A strong example of this principle is Google,[18] and as a consequence, it was able to quickly displace other search engines when it launched its own. The principle also explains why many innovations touted as "the next big thing" quickly flop upon hitting the market. To claim success, an innovation must continuously gain market share, in either adoption or actual profit terms.

Within agriculture, the fast adoption of biotech crops in countries such as Argentina, Australia, Bolivia, Brazil, Canada, China, India, Mexico, Paraguay, South Africa, Uruguay, and the United States, among others, illustrates this concept as well: between 1996 and 2017, their global acreage grew over 100X, from 1.7 to nearly 190 million hectares, out of which 53% were planted in developing countries (ISAAA 2017). This fast adoption story relates to the ability of biotech crops to deliver features which strongly resonate with the pain farmers face in their daily lives at the farm, such as convenience, simplicity, flexibility, speed, and peace of mind in terms of controlling weeds and insects.

An additional way to reduce failure is to launch innovations that only require small or negligible behavior changes, a tactic successfully deployed by Toyota. Its hybrid electric vehicles (e.g., Prius) deliver a driving experience nearly identical to

[11] BMI (body mass index) is a person's weight in kilograms divided by the square of height in meters. A high BMI can be an indicator of high body fat, although not always, and can be used to screen for weight categories that may lead to health issues.

[12] When launched in 1998, the web browser Google provided its users with much more accurate results than existing search engines such as Altavista and Ask and faster results than Yahoo.

that of a gasoline-only car. Consumers didn't need to make a major shift in their behavior to shift to this new product.

1.8.3 Managing Failure

Above all, organizations need to resist doing nothing at all – never implementing new or creative ideas. Robert Sutton of Stanford University suggests that inaction is far worse than failure. Failure, after all, implies some sort of output. Because the quality of innovation is intrinsically linked to quantity of ideas, it makes sense to employ metrics based on quantity of ideas.

Examples of such metrics include how many prototypes built, patents filed, papers published, projects completed, etc. Without a high quantity of attempts, there can be no innovation. Therefore, output – regardless of success or failure – must be rewarded. This may seem unrealistic without additional information.

So how do you reward failure?

According to Sutton, organizations can do the following:

- Make sure people are aware that failure to execute new ideas is their greatest failure and that it will bring about consequences.
- Make certain everyone learns from past failures; do not repeatedly reward the same mistakes.
- If people show low failure rates, be suspicious. Perhaps they are not taking enough risks, or maybe they are hiding their mistakes, rather than allowing others in the organization to learn from them.
- Hire people who have had intelligent failures where lessons have been extracted that enabled subsequent success. Let others in the organization know that's one reason they were hired.

In the pharmaceutical industry, although spending on research and development went up more than 300% industrywide during the 1990s, the number of new drugs approved by the US Food and Drug Administration (FDA) during that period dropped by 50%. Companies see drugs that have cost hundreds of millions of dollars in the R&D phase simply stall out or fail in the FDA approval process. If this attrition does not improve, drug development will become prohibitively expensive.

Vertex Pharmaceuticals is one example of a company that on purpose tries ideas that fail in pursuit of solutions. Vertex uses biotechnology to create transformative medicines targeting diseases such as cystic fibrosis, Duchenne muscular dystrophy, and hemoglobinopathies, among others, and also pain. Paradoxically, Vertex has attempted to decrease the attrition rate by increasing it during the early stages of the innovation process. The driving idea is that the more ideas (molecular combinations) researchers can test, the better their chance of finding a few good ones. Consequently, many ideas lead to dead ends, but at a much earlier stage. Good leads – in this case, effective, safe drugs – have a better chance of getting approved and ultimately generating sales.

Vertex has purposefully placed itself at an intersection of disciplines and technologies where it generates thousands of new drug candidates every day. The goal is to produce as many potential drug compounds as possible to find the intersection of the few combinations that will provide Vertex with a breakthrough drug.

This is how it works in the Vertex example: supercomputers randomly combine molecules with different drug targets. They then throw out the combinations deemed to be ineffective. The remaining molecule combinations go to a team of computer scientists, biologists, chemists, medical doctors, manufacturers, and lawyers, who evaluate them and bring those with the highest potential to fruition. Some compounds are discarded quickly and others a bit later. Some get developed into drugs.

On any given day, Vertex's computers generate thousands of combinations; the vast majority of these end up as failures. By increasing the output of ideas and being unafraid to fail often and quickly, Vertex also has a better chance of developing successful drugs. The strategy is working – at the time of writing, Vertex has already secured FDA approval for four of its products, and several more are in various phases of the approval process for human use.

1.9 The Theory of Jobs to Be Done

One of the main reasons innovations fall short of adoption and value creation is the deep lack of understanding about what drives a user to choose one good or service over another. While marketing departments allocate talented people and significant budgets to reaching end-users, in many cases "the customer rarely buys what the company thinks it is selling him."[19]

The jobs to be done (JTBD thereafter) theory[20] addresses why people buy or decline to buy what companies try to sell them. One of this theory's main advantages is that it dramatically increases the ability to predict the likelihood of innovation success, a welcome addition to the toolbox of innovators. It was developed by the late Clayton M. Christensen and colleagues at Harvard University.

At the core of JTBD is the tenet that users do not purchase goods or services. *Instead, people bring things into their lives to get a specific job done or achieve progress* toward a particular goal under specific circumstances. The goods and services that users acquire are really unconscious means to filling in satisfaction gaps. Unless we understand the "jobs" that people want to fill, the likelihood of innovation success is severely undermined *regardless* of budget size or innovator skill.

A "job" can only exist within a given context – the where, when, who, or what. However, aspects such as the user's life-stage, family status, and financial status also need to be considered. These insights are powerful predictors of user behavior.

[13] A quote from the late author Peter Drucker, known for his work on business management.

[14] Developed and popularized by Clayton Christensen and colleagues at Harvard University, and by Anthony Ulwick from the consulting firm Strategyn.

For instance, "I am short of money" is too vague to allow innovators to fully grasp the "why" at stake. When recast in the framework of a JTBD, however, innovators gain valuable clues needed to predict user behavior: "I am short of money to pay my mortgage which is a source of concern to me. I am a proud provider for my wife and young kids." That kind of insight helps the innovation better meet the needs than the alternatives offered by competitors.

JTBDs can be quite different from solutions. For instance, many providers of MP3 reproducers focused on the presumed solution, namely, providing music. In contrast, Apple aimed the iPod at helping customers listen to music through a seamless experience. It reconsidered the whole business around personal music management, enabling customers to acquire, organize, listen to, and share music.

As value creation targets JTBDs, organizations and firms can not only improve what they already have but also target new, "blue ocean" markets.[21]

The deepest and most significant advantage of this framework is that it explicitly acknowledges that each JTBD consists of several dimensions (see Box 1.5).

Box 1.5 The Anatomy of a Job to Be Done

To fully grasp the *why* behind purchasing decisions, you must understand the complex nature of people. Every effective job must account for all three of these components:

- *Functional* – it addresses practical, technical aspects.
- *Social* – it addresses behavioral aspects of users, as well as how the user interacts and relates to people and groups of people.
- *Emotional* – it addresses very deep, personal aspects of users such as feelings, sentiments, aspirations, and desires.

Think of the job to be done by a luxury Swiss watch:

- *Functional* component – it provides an exact, accurate, and reliable account of the passing of time.
- *Social* component – it conveys an aura of status and economic success, providing cues about social status and power. "I am one of the few who can afford this very expensive and exclusive watch; therefore, I am successful and powerful."
- *Emotional* component – it represents, in this case, a deep sense of belonging to family, passing the watch on from generation to generation. It can also represent the deep act of love from a spouse or partner reflected in such an expensive gift.

---

[15] Describing the concept of blue ocean markets is beyond the scope of this chapter. Interested readers are kindly invited to read the book *Blue Ocean Strategy: How to Create Uncontested Market Space and Make the Competition Irrelevant* by W. Chan Kim and Renee A. Mauborgne.

Most innovation efforts are mainly driven by their functional components. That immediately reduces their likelihood of success, since it is the social and emotional components that are the ones representing a strong, if not the strongest, driver of user preferences and subsequent adoption when people are presented with many alternatives. Within the agricultural domain, innovations solely focused on productivity or technological progress, but which fail to account for cultural and social aspects,[22] are also likely to fall short of expectations.

What are some examples of successfully meeting or failing to address the JTBD?
The failure of many innovations, including those on the cutting edge of technology, can invariably be traced back to their inability to address a JTBD for a large enough constituency. The Segway two-wheeled vehicle is a good example. This cutting-edge technology was introduced to great fanfare and included the strong financial support of very successful entrepreneurs such as Jeff Bezos of Amazon. Yet, it failed to receive widespread adoption. Its failure was due, at least in part, to not enough users truly *needing* it.

In contrast, understanding the *why* behind consumer buying decisions increases likelihood of success. Beat headphones have captured a substantial market share in the United States, even though they are a technically inferior product when compared to headphones produced by makers such as Bose or JBL. How can this be? Beat Electronics appeals to the deep craving for high status of most teenagers. In part due to the association with rapper Jay-Z, users of Beat headphones are seen as "the cool kids" in high school (Wunker and Farber 2019).

The effectiveness of addressing a yet-to-be articulated JTBD is evident in the exponential growth of the mobile phone application Snapchat. It reached 190 million daily active users in 2019 despite fierce competition. The customer base of Snapchat users is much younger than that of other social media. What is so unique about Snapchat? Its main differentiating feature is that messages only last for a few seconds on the recipient's phone.

Why has Snapchat been so successful to date? The JTBD theory explains it: at the time of its launch, Snapchat successfully addressed the emotional component of its JTBD by addressing the needs of teenagers who have long avoided parental scrutiny.

More recently though, the innovations launched by Snapchat have fallen short of addressing the evolving user needs, and therefore its customer base has been eroded, with many of its users deciding to choose competing products such as Instagram.

Finally, a further advantage of the JTBD theory is that it enables organizations to better understand the competitive landscape, to gain insight about previously unnoticed competitors, and to understand the main drivers behind the *why* in the customer's mindset, whether it be social or emotional.[23]

---

[16] These can include gender norms or dietary preferences, depending on the market and country being served.

[17] Amazon does not only compete with other providers of entertainment sharing the same technological means, namely streaming, such as Netflix or HBO. It also competes with any other unrelated provider of leisure and pleasure in the privacy of your own home.

1.10 Business Models

A business model represents how an organization creates, distributes, and captures value, and it is an oftentimes neglected aspect of many innovation efforts, which tend to focus on the good or service to be launched, instead of on how to manage the value created. In addition, as already mentioned, it is unfortunate that the business world tends to be associated only with profit-seeking endeavors. Even innovation teams with enough business acumen oftentimes spend insufficient time and resources on their business models.

The self-inflicted damage caused by organizations paying insufficient attention to designing the right business model not only reduces an innovation's likelihood of success, but it also grants the upper hand to competitors.

A business model is not a business plan. Nor is it a projection of future cash flows, profits, and break-even points. A business model embodies how the organization views value creation, how its goods or services address the JTBD, and how it allocates its talent and resources.

The ability to design business models tests whether an organization is truly customer-centered. In many cases, organizations only pay lip service to keeping customer needs at the center of everything.

Innovative organizations cannot operate without a business model. A well-crafted model lets the organization test hypotheses, formulate and address questions, and challenge underlying assumptions.

A business model can be a mighty source of innovation. Even for companies with a strong digital footprint, a well-designed business model is key to transformational growth (Johnson 2018). Business models can be nicely articulated with the *canvas framework* developed by Osterwalder and Pigneur (2010). A canvas is a visual chart which is very effective to describe the founding blocks of the business model of an organization, good, or service. Ulvenblad provides in Chap. 5 of this book a detailed account about business models in Sweden.

With the right adaptive business model in place, an organization can rapidly evolve and achieve success.

1.10.1 The Components of Business Models

A business model can be articulated around four components (Fig. 1.2):

Taken together, these four components can yield a powerful new way of doing business and achieving innovation.

1.10.1.1 The Value Proposition

Value propositions are closely related to the JTBD concept. *The value proposition is the promise made to users.* A value proposition is a good or service enabling users to carry out a JTBD in a more convenient, effective, and/or affordable way (Eyting

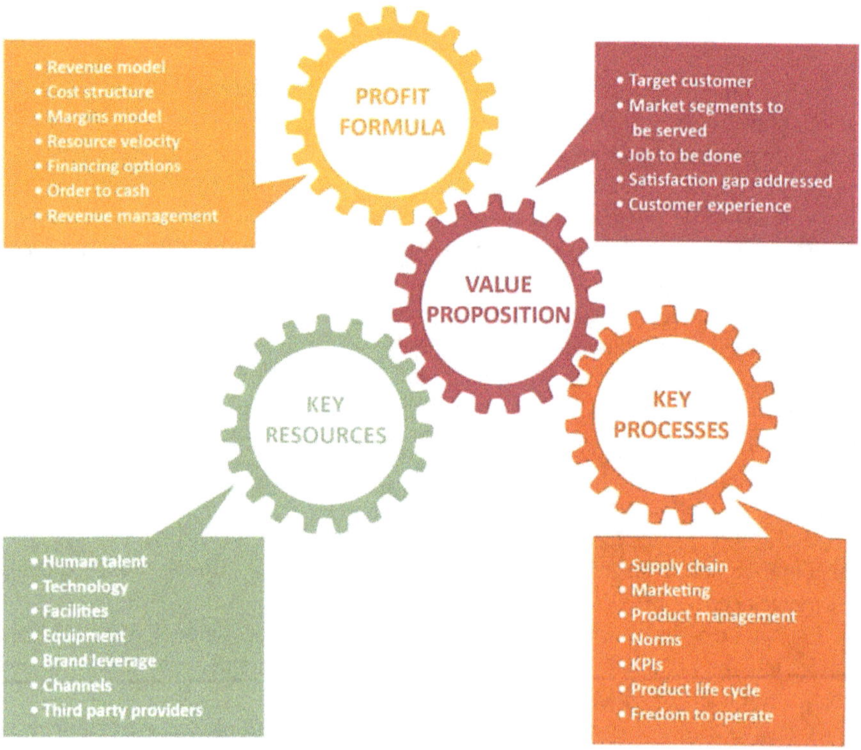

Fig. 1.2 The four components of a business model. (Adapted from Johnson 2010)

et al. 2011). It fulfills the job better than all other alternatives and at an appropriate price (Boxes 1.6 and 1.7).

While designing a value proposition, questions from the perspective of the customer – not from the organization – should be addressed:

- Is my JTBD properly and truly addressed?
- What will the good, the service, or a combination of them look like?
- Who is giving me this offer?
- How are they giving it to me?
- If it is a product or good, how do I dispose of it?
- What tradeoffs are imposed on me along with the offering?
- What does the landscape of payment options look like?

Once these questions are answered, a value proposition can be developed through experiments to better understand the JTBD.

Ultimately, developing the value proposition is by far the most important component of any business model. Any organization wanting to innovate and succeed must spend time on developing the right value proposition before moving forward.

Box 1.6 Hortifrut: Worldwide Leader in Berries Production

Hortifrut is a Chilean company and a global leader in the production and commercialization of blueberries. As the world's second largest provider of berries, it enjoys a comfortable 25% of global market share of blueberries and has sales in over 35 countries. Its value proposition can be summarized as "berries for the world – every day." That appeals to both wholesale customers and retailers such as Carrefour, Walmart, and Costco, as it simplifies and strengthens supply chains. At the same time, this value proposition enables the company to capture attractive prices all year instead of facing periods when prices drop because of excessive supply.

Box 1.7 Safaricom and M-Pesa: The Power of an Effective Value Proposition

While experimenting with diverse business models to develop a micro-loan business in Kenya, the telecom firm Safaricom found a much more compelling value proposition: a cheap, safe, and reliable system to send money by mobile phone to friends and family in rural areas. This insight became the key value proposition of M-Pesa (M for Mobile, Pesa for money in Swahili). M-Pesa is a branchless banking system that allows customers to pay bills, deposit, withdraw, transfer, and save money in a virtual account. In some markets, like Kenya, users can use a mobile phone to transfer money between the service and a bank account. As of 2018, Safaricom reported more than 28 million M-Pesa users and over 155,000 M-Pesa agents in Kenya. There are over 10 million active users in countries other than Kenya.

From its humble beginnings in 2007, M-Pesa has become a huge economic player in Kenya: it processes over 1.7 billion transactions a year, accounting for more than 50% of Kenya's GDP value.

1.10.1.2 Profit Formula

The profit formula articulates how an organization creates value for itself, and for its shareholders/stakeholders. Its main components are (1) the revenue model, (2) the cost structure, (3) the margins model, and (4) resource velocity (Johnson et al. 2008). Revenue models establish the price at a given volume of transactions to cover overhead, fixed costs, and desired profit margins. Profit formulas also establish the resource velocity at which the organization turns over assets to achieve adequate returns. Profit formulas need to account for the customer's willingness to pay.

Do not assume that business models and profit formulas are interchangeable. The profit formula is *an element of* a business model.

The value proposition defines value for the customer, whereas the profit formula establishes value for the organization and its shareholders/stakeholders (Johnson 2010).

1.10.1.3 Key Resources

Key resources are competencies such as human talent, technology, facilities, equipment, channels, reputation, freedom to operate, and brand. They deliver in an effective way the value proposition to end-users. Key elements create value for the user and the organization. An organization needs to understand how those elements interact with each other. Generic elements found in every organization and which do not create competitive differentiation are not considered key resources (Johnson et al. 2008). For retailers such as Amazon, a key resource is its Information Technology infrastructure. Within the agricultural domain, examples of key resources can include the scientific expertise, market understanding, and proprietary knowledge at companies such as Bayer, Benson Hill, Calyxt, Cargill, Corteva, Enko, HZPC, Inari, Indigo, John Deere, KWS, Mosaic, Pairwise, Provivi and Syngenta among others, as their competitors may lack such resources. By leveraging such key resources, these companies increase the value and benefits their users get from the goods and services they offer.

1.10.1.4 Key Processes

Key processes may include training, budgeting, manufacturing, planning, sales, and services. Such processes enable companies to deliver value propositions to users on a recurrent basis which is then able to be scaled up (Johnson 2010). In the case of Hortifrut, a key process is its ability to supply high-quality, fresh berries to customer throughout the whole year because it sources from production fields in different environments, including Chile, Peru, Spain, and Mexico. Building on previous expertise and local partners, it is also producing and commercializing berries in China. In the case of Tuskys, one of the leading supermarket chains in Kenya, which commercializes bread and pastries enriched with vitamin A, a key process is the method of incorporating the puree of vitamin A-enriched sweetpotato into their proprietary bakery operations.

1.11 Closing Remarks

This chapter lays the foundation for successful innovation. These concepts can be applied to any economic sector – to profit-seeking and nonprofit organizations, and to both private and public endeavors. If there is one economic sector with an unparalleled application potential, however, it is agriculture and agrifood systems in both developed and developing countries. The ultimate goal of all innovation endeavors in this sector should be improving the quality of life, the outlook, and the well-being of farmers and end-users.

Firms seeking profits and nonprofit organizations remain as absolutely valid innovation pursuits. But such pursuits should be seen as a means to that end. Every innovation effort must allocate as much human talent and as many resources in developing business models as they spend on developing solutions which address the JTBD for end-users.

Taken together, the insights, concepts, and examples provided in this chapter articulate the pathway and the way forward to increased success and deeper impact of innovation efforts, which are urgently required in agriculture and agrifood systems.

Acknowledgments I would like to sincerely thank Oscar Ortiz and Graham Thiele from the International Potato Center for thoughtful comments on a draft version of this chapter. Thanks also must go to Noelia Campos, Catherine Scott and David Poulson for their review of this chapter, much improving its style and readability. Figures 1.1 and 1.2 are adapted and reprinted with permission from Harvard Business Publishing. Figure 1.1 from "What is Disruptive Innovation" by Clayton M. Christensen, Michael Raynor and Rory McDonald, Harvard Business Review December 2015. Harvard Business Press 2015 by Clayton M. Christensen, Michael Raynor and Rory McDonald, all rights reserved. Figure 1.2 from "Seizing the White Space: Business Model Innovation for Growth and Renewal" by Mark W. Johnson. Harvard Business Press 2010 by Mark W. Johnson, all rights reserved.

References

Autor D (2014) Skills, education, and the rise of earnings inequality among the "other 99 percent.". Science 344:675–681

Anthony S, Viguerie P, Schwartz E, Van Landeghem J (2018) 2018 Corporate longevity forecast: Creative destruction is accelerating. Available at https://www.innosight.com/wp-content/uploads/2017/11/Innosight-Corporate-Longevity-2018.pdf [verified on November 4th, 2019]

Bason C, Austin R (2019) The right way to lead design thinking. Harv Bus Rev 98:1–11

Christensen C, Raynor M, McDonald R (2015) What is disruptive innovation? Harv Bus Rev 94:1–11

Christensen C, Ojomo E, van Bever D (2017) Africa's new generation of innovators. Harv Bus Rev 96:245–255

Dillon K (2020) Disruption 2020: an interview with Clayton M. Christensen. MIT Sloan Management Review. Spring 2020:2–7

Eyting M, Johnson M, Nair H (2011) New emerging business models in emerging markets. Harv Bus Rev 90:1–9

Frederik R (2015) Making innovation work: ambidextrous organizations in the seniors housing and care industry. Seniors Hous Care J 23:76–84

Gourville J (2006) Eager sellers and stony buyers: Understanding the psychology of new products adoption. Harv Bus Rev 84(6):98–106

Govindarajan V (2016) The three box solution: A strategy for leading innovation. Harvard Business Press, Boston

ISAAA (2017) Global status of commercialized biotech/GM crops in 2017: biotech crop adoption surges as economic benefits accumulate in 22 years, ISAAA Brief No. 53. ISAAA, Ithaca

Jayne T, Yeboah F, Henry C (2017) The future of work in African agriculture: trends and drivers of change. Working Paper No. 25. International Labor Organization. Available at: https://www.ilo.org/wcmsp5/groups/public/%2D%2D-dgreports/%2D%2D-inst/documents/publication/wcms_624872.pdf. 11 Oct 2019

Johnson M (2010) Seizing the white space. Harvard Business Press, Boston, 208 p

Johnson M (2018) Reinvent your business model: How to seize the white space for transformative growth. Harvard Business Press, Boston

Johnson MW, Christensen CM, Kagermann H (2008) Reinventing your business model. Harv Bus Rev 87:52–60

Kahneman D, Tversky A (1979) Prospect theory: an analysis of decision under risk. Econometrica 47:263–291

Kahneman D, Knetsch J, Thaler R (1991) Anomalies: the endowment effect, loss aversion, and status quo bias. J Econ Perspect 5(1):193–206

Karl F, Holle R, Schwettmann L, Peters A, Laxy M (2019) Status quo bias and health behavior: findings from a cross-sectional study. Eur J Public Health. https://doi.org/10.1093/eurpub/ckz017. 11 Oct 2019

Liedtka J (2018) Why design thinking works. Harv Bus Rev 97:2–9

Liedtka J, Salzman R, Azer D (2017) Design thinking for the greater good. Columbia University Press, New York, 342 p

Low JW, Mwanga ROM, Andrade M, Carey E, Ball AM (2017) Tackling vitamin A deficiency with biofortified sweetpotato in sub-Saharan Africa. Glob Food Sec 14:23–30

Macchiavelli N (1992) The prince. Dover Thrift Editions. 72 p

Markovitch DG, O'Connor CG, Harper PJ (2015) Beyond invention: the additive impact of incubation capabilities to firm value. R D Manag 47(3):352–367

Organisation for Economic Cooperation and Development (OECD) (2015) Frascati manual 2015 #x2014;Guidelines for collecting and reporting data on research and experimental development (FM 7.0). Paris

O'Reilly C, Tushman L (2004) The ambidextrous organization. Harv Bus Rev 83:84 71

Osterwalder A, Pigneur Y (2010) Business model generation. Wiley, Hoboken

Rosellon M, Del Prado F (2017) Achieving innovation without formal R&D: Philippine case study of garment firms. Philippine Institute for Development Studies. Discussion Paper No. DP 2017-09

Schumpeter JA (1934) The theory of economic development: an inquiry into profits, capital, credit, interest, and the business cycle. Transaction Books, New Brunswick

Schumpeter JA (1942) Capitalism, socialism, and democracy, 3rd edn. Harper& Row, New York

Taneja N (2017) Airline industry: poised for disruptive innovation? Routledge Press, Abingdon

Viki T (2016) Why R&D spending is not a measure of innovation. Forbes. August 2016 issue

Ward P, Singh V (2014) Risk and ambiguity preferences and the adoption of new agricultural technologies: evidence from field experiments in rural India. IFPRI Discussion Paper 1324. https://doi.org/10.2139/ssrn.2392762. 11 Oct 2019

Wunker S, Farber D (2019) A winning formula: disruptive innovation + jobs to be done. Rotman Management Magazine. Winter 2018 issue

WWDC. 1997. Steve jobs about Apple's future. http://www.youtube.com/watch?v=qyd0tPOSK6o. [verified on Oct 11th 2019]

Yoo Y, Kim K (2015) How Samsung became a design powerhouse. Harv Bus Rev 94:1–8

2

Innovation, Agricultural Productivity and Sustainability

Ann Steensland and Margaret Zeigler

2.1 The Global Agricultural Imperative

In 2050, the number of people living on our planet will grow to nearly 10 billion, and that could double the demand for food, feed, fiber, and biofuels from 2005 levels (von Lampe et al. 2014). It is imperative that this demand be met in a way that is economically viable, environmentally sustainable, and socially beneficial.

Our food and agriculture systems face enormous challenges to sustainably producing sufficient, nutritious affordable food, feed, fiber, and biofuel for a growing world. At present, agriculture is the largest user of water globally; agriculture also is the single largest use of land, covering a third of the planet's surface. Competition between food production and other uses of water and land will increase in the coming decades. In addition, climate change threatens agricultural productivity due to increased temperatures and shifts in weather patterns (Box 2.1), thereby making it difficult for crops and livestock to grow and thrive and for agricultural laborers to endure the physical challenges.

Box 2.1 The Challenge of Climate Change for India's Farmers (Naresh et al. 2017)

India's farmers are struggling as temperature and rainfall patterns become hotter, drier, and wetter. By the end of the century, the mean summer temperature in India could increase by five degrees Celsius. The number of days of extreme heat could increase by more than a month, and the number of warm nights could more than double. The amount of rain is also

(continued)

A. Steensland
Global Agricultural Productivity Initiative, Virginia Tech, Blacksburg, VA, USA
e-mail: anns@vt.edu; https://globalagriculturalproductivity.org/

M. Zeigler (✉)
HarvestLAC, Washington, DC, USA

(continued)

expected to increase by as much as 40%, while the frequency of extreme rain events is also increasing, as well as the number and length of droughts.

Under these conditions, by 2035, yields for India's major food crops are

expected to decline by as much as 10%. Rising temperatures and the increase in extreme heat will make living and working conditions unbearable and reduce the productivity of farmers and agricultural laborers. Livestock will also struggle with the heat, and nutrition of their fodder will be reduced.

Without support and adaptation, agricultural productivity in India could decline by as much as 25%; the productivity of small-scale rain-fed farms could decline by as much as 50%, posing formidable challenges to food security, human well-being, and economic and political stability.

Volatile agricultural business cycles also create challenges for farmers as they seek to manage risk and invest for the future. Conflict and migration generate famine and human suffering. And global health is compromised by malnutrition, poor diets, and disease.

The previous 10 years have witnessed unprecedented demand for agricultural commodities, driven by income increases and population growth in China and India, as well as demand for biofuels stimulated by high energy prices.

Over the decade 2017–2026, the OECD and the United Nations Food and Agriculture Organization (FAO) project that the rate of demand growth for all agricultural commodities will slow compared with the prior decade (OECD/FAO 2017). The rate of demand growth for cereal grains, meat, fish, and vegetable oil will be cut nearly in half, the notable exception being increasing demand for fresh dairy (Box 2.2 and Fig. 2.1). OECD and FAO attribute the decline in the rate of demand growth to moderating rates of economic growth, particularly in China, and a decline in demand for biofuels.

While the rate of demand growth may be slowing (compared to the previous 10 years), the overall demand for food and agriculture products is still rising, as is the global population. In fact, the highest demand growth for many agricultural

Box 2.2 Meeting India's Milk Demand (Steensland and Zeigler 2018)

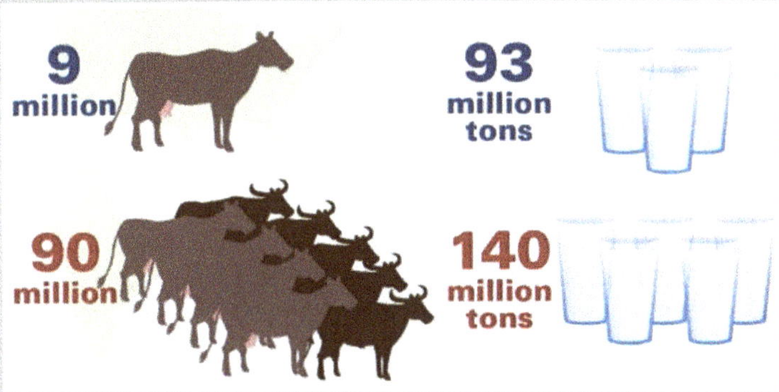

Fig. 2.1 Milk productivity in India and the United States. India has ten times as many dairy-producing bovines (cattle and buffalo) as the United States but produces only 50% more milk. (Data from FAOSTAT 2014). Figure: adapted from 2017 Global Agricultural Productivity Report (GAP Report), page 14)

Over the next decade, India will account for 54% of the increase in global demand for fresh dairy products, requiring an additional 56 million tons of milk. India is already the largest dairy producer in the world, but dairy cattle and buffalo productivity is low. In 2014, India had 50 million dairy cows and 40 million water buffalo, a total of 90 million animals producing 140 million tons of milk. Dairy cattle produce an average of 14,000 hectograms per animal, and buffalo produce 19,000 hectograms per animal (FAOSTAT 2014).

By contrast, the United States had just 9.2 million dairy cows and produced more than 93 million tons of milk, an average of 101,000 hectograms per animal (OECD/FAO 2017). Given the projected demand in India, improving the health and productivity of the current dairy cow and buffalo populations needs to be prioritized. Indian farmers and consumers are increasingly choosing buffalo over dairy cow milk (Landes et al. 2017). Consumers prefer the higher fat content of buffalo milk, and it brings a higher return to farmers. Buffalo are more adaptable to the changing climate in India, and they convert the low-quality indigenous grasses into milk more efficiently than cattle.

Improving genetics, feed, and animal care practices can provide more milk using fewer animals. Increasing access to mechanization for small- and medium-scale farmers would reduce reliance on cattle for draught power, allowing investments in milk production.

products is coming from regions most vulnerable to climate change, with high rates of population growth and low rates of agricultural productivity, such as South Asia and sub-Saharan Africa. These regions are characterized by small farms, with little access to productive inputs, and a substantial proportion of the rural workforce represented by women. As production increases to meet the growing demand, concerns are rising about the environmental impact these low-productivity systems will have on the natural resource base, along with rising greenhouse gas emissions.

2.2 What Is Productivity in Agriculture?

For agricultural producers of all scales, there are multiple approaches to meeting the current and future demand for agricultural products:

- *Land Expansion* – Producers use more land to produce more agricultural products and, in some cases, convert forest to cropland or rangeland.
- *Irrigation* – Producers deploy or extend irrigation systems to protect land against drought and improve its productive capacity, which may permit multiple cropping seasons. If not carefully managed, groundwater may be depleted.
- *Intensification* – Producers increase applications of fertilizer, machinery, labor, seeds, herbicides, or other inputs on existing land to grow more crops or raise more livestock.

Meeting demand in a way that reflects the needs of producers and consumers today, while safeguarding future agricultural capacity, is best achieved another way:

- *Productivity Growth* – Adopting technologies and production practices that result in more output from the same amount, or less, inputs. This can be measured as *total factor productivity (TFP).*

While the terms are sometimes used interchangeably, agricultural "productivity" is distinct from "output" and "yield." Output is the gross amount produced, and yield measures the amount of output per unit of production, usually land. TFP (Fig. 2.2) is the ratio of agricultural outputs (gross crop and livestock output) to inputs (land, labor, fertilizer, feed, machinery, and livestock). TFP measures changes in the efficiency with which these inputs are transformed into outputs.

TFP is calculated using measurable inputs, so water and seeds are not factors in the equation. Eighty percent of global agriculture is rain-fed, making it difficult to quantify water usage. It is also difficult to quantify seed usage since millions of farmers, particularly those at smaller scales, use open pollinated varieties (OPVs), which are derived from the grain of the previous harvest.

By measuring TFP, as opposed to yields or output, we begin to understand the extent to which increased output is due to better use of these critical resources. Policymakers, development agencies, researchers, and producers use TFP to identify where improvements are needed in agricultural production systems and to determine which investments and policies increase productivity and enhance sustainability.

Producers, governments, and agribusinesses who pursue this course are not just interested in whether agricultural output is growing but to what extent increased output is due to better use of existing resources through the application of improved products, technologies, and practices – *essentially, how innovative their operations are.*

Fig. 2.2 Total factor productivity. (Source: 2018 Global Agricultural Productivity Report (GAP Report))

Examining TFP is the best way to get that information, which can be enormously useful in identifying where improvements are needed in agricultural production systems, how to make investment decisions, and what policies support more productive and sustainable agriculture.

2.3 Productivity and Innovation in Practice

For crops, improved TFP results from adopting innovations like higher-yielding, pest-resistant, and/or drought- and flood-tolerant seed varieties. The growing bio-innovation sector includes precision use of microbes (bacteria and fungi) to help crop farmers generate more yield on the same land. Microbes also protect plants from dry conditions and increase yield, as well as protect plants from pests.

Agricultural extension agents or agricultural retail service providers can equip growers with knowledge of best practices that enable more efficient and timely cultivation techniques, improve soil and water quality, and improve crop yields. TFP growth also comes from widespread adoption of precision data and information technologies in farm equipment to target applications of fertilizer, water, and crop protection. Having access to geo-referenced data also enables farmers to improve soil quality, plan for crop rotation cycles, and place less productive land into conservation.

In livestock production, TFP increases when favorable genetic traits in animals are selected and bred and when animals receive better overall husbandry, vaccinations, and high-quality feeds that deliver more nutrition per volume. In forestry, genetically improved trees provide faster-growing products for earlier harvesting and more volume per tree.

Ensuring that farmers and producers of all scales and sizes gain access to better innovation, technology and training, and knowledge for best practices will help foster greater TFP and reduce impact on the soil, aquifers, and other underground water bodies and water and air quality, as well as effectively use increasingly scarce labor in agricultural operations.

TFP looks beyond simply *how much* farmers are producing. It reveals *how efficiently* they are producing it and indicates how well they are conserving available resources to meet future needs. Productivity growth in agriculture lowers the cost per unit of output, helping producers succeed in today's competitive business cycle, and enables agri-food systems to provide lower prices for consumers.

Farmers use productive technologies and practices such as improved seeds and farm equipment, genetically improved livestock, and good animal husbandry to increase output while conserving land and water and protecting soils for future generations. In addition to promoting competitiveness and conservation, productive technologies and good practices also support the *UN Sustainable Development Goals (SDGs)* to end hunger and malnutrition, protect the safety of the water supply, and reduce greenhouse gas emissions.

Case studies throughout this chapter demonstrate how farmers of all scales, producing a variety of products in different geographies, are conserving and protecting their soil and water resources while reducing their climate impact. Innovations highlighted include drought-tolerant new plant varieties that enable poor farmers in dryland areas to grow in stressful conditions (Box 2.3); precision agriculture technologies that enrich soil in the field and keep nutrients out of streams; and animal care innovations and practices that improve the health and productivity of each animal while reducing emissions from livestock production.

Box 2.3 Corn Productivity Feeds Vietnam (Zeigler and Steensland 2017)
Corn is already the second largest crop in Vietnam after rice, yet the country still imports between five and seven million tons of corn each year to feed livestock for growing consumer protein demand. With little additional land available for production, farmers must improve corn productivity on existing

(continued)

(continued)
land to seize the market opportunity to supply livestock feed. Better production practices and better seeds are needed.

As part of Monsanto (now Bayer AG) Vietnam's sustainable development efforts, more than 200,000 farmers have received training since 2015 on good agronomic practices and hybrid corn seed selection. Seeds with beneficial traits improved through conventional breeding and advanced biotechnology are now becoming available for many farmers in Vietnam.

Vietnamese farmer, Huynh Van Hue, enjoys a successful corn harvest using the rice-to-corn rotation protocol.

High-yielding improved corn seeds such as hybrids and stacked trait biotechnology (seeds engineered to deliver to farmers both insect protection and herbicide tolerance traits) help farmers grow more while requiring less labor to remove weeds and apply crop protection. The improved corn is particularly resilient against three harmful pests: Asian corn borer, common cutworm, and corn earworm.

To help farmers make the transition from rice to corn, Monsanto agronomists and rice farmers developed a series of best agronomic practices, the *Dekalb® Cultivation Rice-to-Corn Rotation Protocol*, that was selected as a preferred cropping system by the *Vietnamese Ministry of Agriculture and Rural Development*. Farmers in pilot programs used this protocol across several departments of Vietnam and increased their incomes by up to 400% while supplying more corn for livestock feed. New jobs and businesses such as corn drying and feed mill development are becoming part of the growing corn value chain.

The Ministry has set a goal of transitioning 668,000 hectares of rice-growing land to corn production in the northern region of Vietnam by 2020. Farmers in other regions of the country are also being supported as they diversify to more resilient, high value crops and livestock while sustainably intensifying rice production in the most suitable areas.

2.4 Productivity Rises, with Room to Grow

TFP accounts for the largest share of growth in global agricultural output today (Fig. 2.3). In the 1960s, the Green Revolution introduced high-yielding new plant varieties of wheat and rice to millions of small farmers in Mexico, India, and other developing countries, along with access to fertilizers, irrigation, and machinery. As farmers began to use those inputs more efficiently, the contribution of inputs per land area to agriculture output declined (orange bar), and TFP's contribution increased (green bar).

Agricultural productivity supports the needs of producers, consumers, and the environment. Productive use of inputs and capital helps farmers control costs during volatile business cycles. Consumers benefit from lower food prices and natural resources, particularly land and water, are conserved.

However, the most recent 10-year period of available data (2006–2015) reveals that TFP's contribution to output growth is declining and more output has been generated by placing additional land into production. (Compare the two columns, 2001–2010 and 2006–2015, in Fig. 2.3.) Farmers around the world expanded their production in response to lower global grain stocks and higher prices during this period.

In high-income countries, improvements in productivity expand output while reducing inputs used in agriculture and dramatically freezing land expansion (Fig. 2.4). Innovations that have raised productivity include advanced crop technologies (genetically modified seeds, novel genetic and breeding approaches, and improved crop protection products) along with advanced livestock breeding, improved animal feed and care, precision agriculture, and better nutrient management. In the most recent decade, however, the downward trend in productivity growth can also be seen in high-income countries. (Compare the two columns, 2001–2010 and 2006–2015, in Fig. 2.4.)

Low-income countries have mirrored the global trend in TFP growth and enjoyed a substantial increase in agricultural output since the 1960s (Fig. 2.5). However, since the 1980s, opening new land for agricultural production (red bar) remains the primary driver of agricultural output. TFP's contribution to agricultural output has grown during the most recent 10-year period. (Compare the two columns on the right in Fig. 2.5.)

Nonetheless, economic and political forces have driven land expansion in low-income countries: transitions to market-based economies, the introduction of input subsidies and prices supports, growing populations needing more land to cultivate, and the extension of irrigation. While some land is suitable for agricultural expansion, greater productivity on existing cultivated land needs to be prioritized to minimize agriculture's impact on soil, water, forests, and wildlife.

Low labor productivity on small-scale farms, predominantly found in low-income country agricultural systems, largely accounts for the higher inputs per hectare of agricultural land results (Fig. 2.5, orange bar). Small-scale farms are labor-intensive due to insufficient off-farm or urban employment opportunities that could absorb the excess labor in rural areas. Small-scale farmers also struggle to

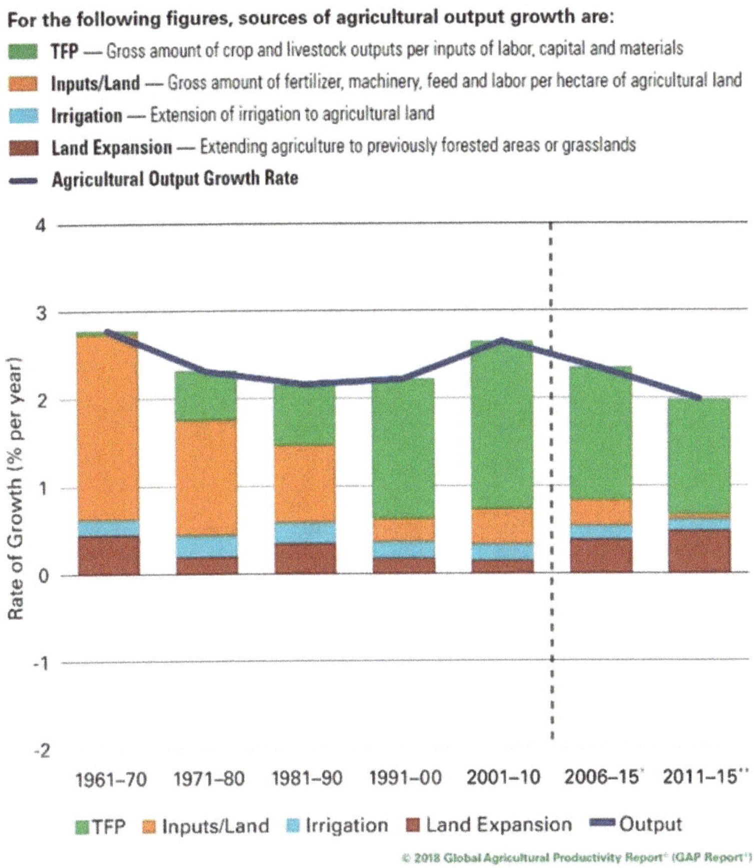

Fig. 2.3 Sources of growth in global agricultural output, 1961–2015. *Depicts data for the most recent ten-year period. **Depicts data for the most recent five-year period. (Source: USDA Economic Research Service (2018))

purchase or rent machinery at competitive prices relative to their labor cost and, in addition, lack the market insight needed to capture better prices for their produce. This contributes to high rates of rural poverty and food insecurity.

2.5 Agricultural Productivity and the Sustainable Development Goals

The United Nations' 17 Sustainable Development Goals (SDGs) took effect at the beginning of 2016, launching the countdown to achieve inclusive, sustainable development and economic growth by 2030. Many SDGs have clear implications for agriculture, while agriculture and forestry play a central role in the strategy to

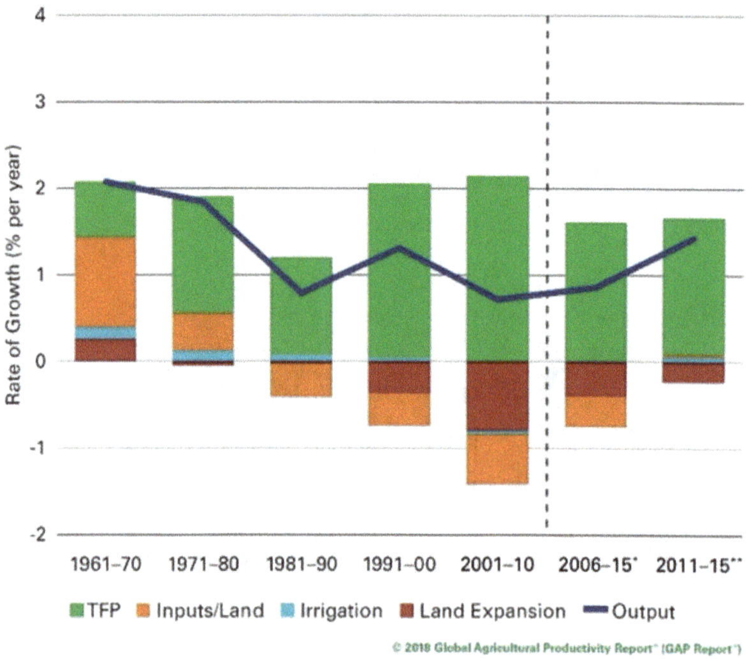

Fig. 2.4 Sources of growth in agricultural output: high-income countries, 1961–2015. *Depicts data for the most recent ten-year period. **Depicts data for the most recent five-year period. (Source: USDA Economic Research Service (2018))

achieve many of the goals. Most notably, *Sustainable Development Goal 2 (SDG 2)* calls the world community to "end hunger, achieve food security and improved nutrition, and promote sustainable agriculture." As part of a comprehensive set of actions, the *UN's 2030 Agenda for Sustainable Development* calls for "doubl[ing] the agricultural productivity and incomes of small-scale food producers, particularly women, indigenous people, family farmers, pastoralists and fishers, including through secure and equal access to land, other productive resources and inputs, knowledge, financial services, markets, and opportunities for value addition and non-farm employment" (von Lampe et al. 2014).

Accelerating agricultural productivity must be at the core of a comprehensive strategy to sustainably feed the world (Box 2.4). With more than three-quarters of the world's poor being heavily dependent on agriculture for their direct subsistence food needs as well as for their incomes, agricultural development through productivity improvements and higher incomes is one of the most powerful ways that farmers, pastoralists, and fishers can rise out of poverty and improve their nutrition and health.

Productivity benefits producers of all sizes by improving the resilience and competitiveness of their operations. Productivity also enables better stewardship of land, water, and other natural resources.

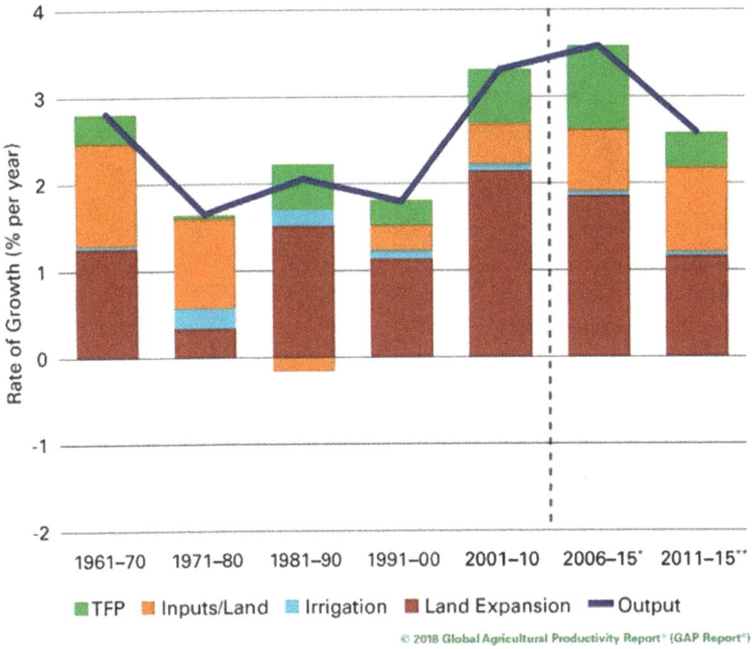

Fig. 2.5 Sources of growth in agricultural output: low-income countries, 1961–2015. *Depicts data for the most recent ten-year period. **Depicts data for the most recent five-year period. (Source: USDA Economic Research Service (2018))

Box 2.4 Doubling Agricultural Productivity Is the Right Goal

The projected slowdown in demand for food and agriculture products over the next decade has prompted calls for a reduction in the agricultural output targets for 2050 (Hunter et al. 2017). Yet a large and growing body of sophisticated modeling by agricultural economists examining long-term scenarios for agriculture, food, and the environment indicates that it may be too soon to consider revising these goals downward.

The Agricultural Model Intercomparison and Improvement Project (AgMIP) is an international collaborative effort to improve agricultural economic models. AgMIP coordinates regional and global assessments of climate impacts and uses multiple scenarios for crop and livestock production across differing geographies to explore the effects of uncertainty, data selection, and methodology on the models' results.

AgMIP's analysis of ten leading global multi-sectoral projection models found that world agricultural production of crops and livestock between 2005 and 2050 will need to rise by between 60% and 111%, with demand growth particularly strong for ruminant products (cows, sheep) as well as for com-

(continued)

(continued)
modities used in the production of biofuels – sugar, coarse grains, and oilseeds (von Lampe et al. 2014). (The OECD/FAO prediction of a decrease in the rate of demand growth for food and agriculture products extends only to 2026, not to 2050.)

Most importantly, AgMIP points to the impact climate change will have on the ability of agriculture to meet future demand. The ten models suggest that climate change will generate higher prices for agricultural commodities in general and particularly for crops (von Lampe et al. 2014). The impact of climate change must be considered to avoid a downward bias in projected supply estimates.

2.6 Tracking Productivity: The GAP Index™

The 2018 Global Agricultural Productivity (GAP) Index™ reveals that for the fifth straight year global agricultural productivity growth (TFP) is not accelerating fast enough to sustainably meet the food, feed, fiber, and biofuel needs of nearly 10 billion people in 2050.

In 2010, the Global Harvest Initiative (GHI) calculated that global agricultural productivity (as measured by TFP) must grow by an average rate of at least 1.75% annually to double all agricultural output *through productivity growth* by 2050. The US Department of Agriculture's Economic Research Service (USDA ERS) estimates that since 2010, TFP growth globally has been rising by an average annual rate of only 1.51% (Fig. 2.6).

The GAP Index™ was created in collaboration with Dr. Keith Fuglie of USDA Economic Research Service. Dr. Fuglie provides annual updates of TFP data for the GAP Report.

The average annual TFP growth rate in low-income countries is particularly troubling. Sustainable Development Goal 2 (SDG 2) calls for doubling productivity for small-scale farmers in the low-income countries. The current annual rate of TFP growth in low-income countries is only 0.96%, down from 1.5% 3 years ago. This is well below the TFP growth rates needed to achieve the SDG 2 target of doubling productivity for small-scale farmers in the lowest-income countries by 2030.

If this trend continues, farmers in low-income, food-deficit countries (where population growth is rapidly rising) will use more land and water to increase their output, straining a natural resource base already threatened by extreme weather events and climate change. Many low-income countries will need to import food but lack sufficient income to purchase enough to meet the needs of their citizens. Poor urban households will bear the brunt of higher food prices in these countries, but they will also impact low-income rural populations since they are net food buyers. Some of the food demand will not be met, and millions of people will be debilitated by hunger and malnutrition.

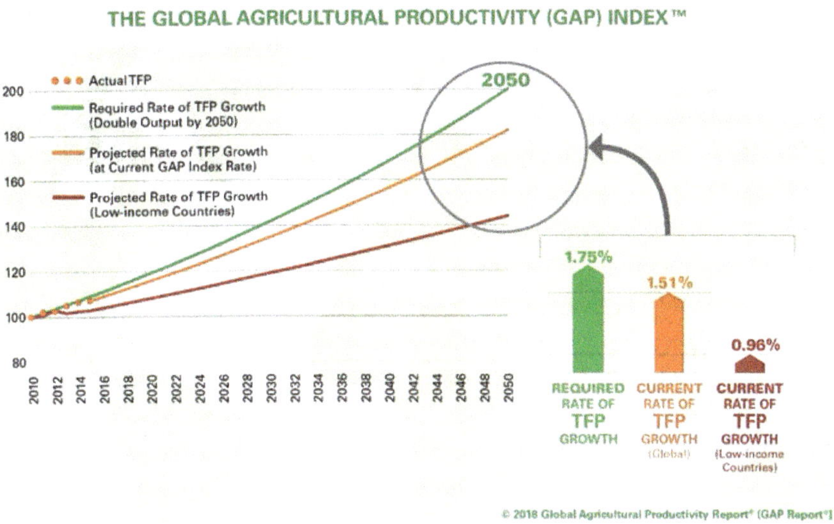

Fig. 2.6 The Global Agricultural Productivity (GAP) Index™ (2018). (Source: Food Demand Index is from Global Harvest Initiative (GHI) (2018); Agricultural Output from TFP Growth is from USDA Economic Research Service (2018))

2.6.1 Regional TFP Growth Rates Raise Concerns

Rates of productivity growth vary greatly by region, as can be seen by comparing food demand indexes against projected agricultural output from TFP growth for the period 2000 to 2030. Figure 2.7 compares the percentage of the estimated food demand for 2030 that can be met with projected TFP growth for six world regions and China.

At current rates of TFP growth, sub-Saharan Africa (SSA) will meet only 8% of its food demand through productivity (Fig. 2.7). Trade plays a key role in closing Africa's food demand gap; 50% of its vegetable oils, 35% of its poultry meat, and 23% of its sugar requirements are imported (OECD/FAO 2016). Without significant increases in agricultural productivity growth, African countries will not meet their SDG targets for reducing hunger, malnutrition, and poverty and will rely more on trade to meet growing demand and most likely will continue to expand land area under cultivation to grow more food, threatening wildlife habitat and releasing soil carbon from forest conversion to cropland. In addition, an increasing financial burden will be imposed on them in order to increase their imports of raw food materials which oftentimes are paid for in foreign currencies.

With 60% of the world's population and considerable economic diversity, the Asian regions (South Asia, Southeast Asia, East Asia, including China) exhibit varying degrees of capacity to meet food demand through productivity.

China has prioritized agricultural development and food security and has achieved great progress in reducing hunger. Yet with little arable land and growing affluence, China will require more investments in productivity and more trade to meet future demand.

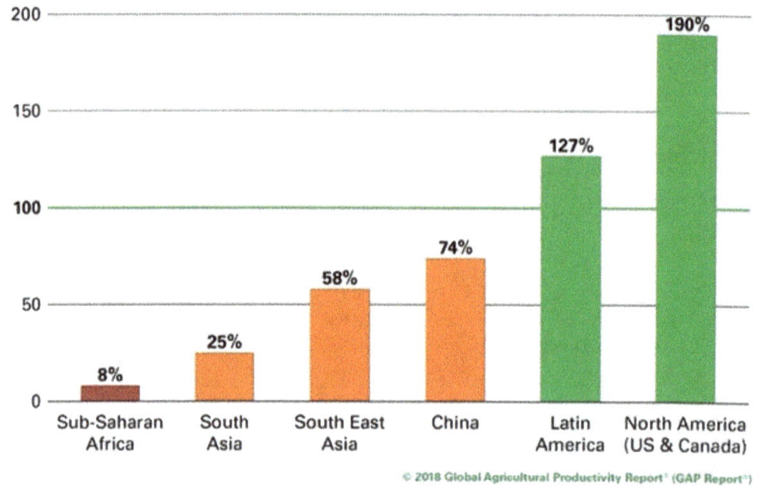

Fig. 2.7 Percent of food demand met through productivity (TFP) growth in 2030. *Note on methodology*: The projection of agricultural output from TFP growth uses USDA ERS (2017) estimates of average TFP growth during 2004–2014 and assumes this is maintained through 2030. The projected growth in food demand uses UN estimates of population, World Bank estimates of GDP forecasts and PricewaterhouseCoopers LLP (PwC) estimates of GDP growth in PPP, and estimates of the income elasticity of food demand from Tweeten and Thompson (2008).The income elasticity of food demand indicates the share of the growth in per capita income that will be spent on food. Multiplying the income elasticity by the growth rate in per capita income gives the growth rate in per capita food consumption holding food prices fixed. Adding this to the population growth gives the total growth in food demand for a given price level. (Source: Food Demand Index is from Global Harvest Initiative (2017). Agricultural Output from TFP is from USDA Economic Research Service (2017))

South Asia will only meet 25% of its growing demand through productivity by 2030. Despite increasing agricultural output since the Green Revolution, India still relies on large amounts of inputs per land area and high labor inputs to produce food, rather than boosting productivity. Other Asian countries, such as Indonesia and Vietnam, could potentially reduce hunger and improve agricultural productivity, but face significant threats from climate change, requiring accelerated investments to keep up with the challenge.

Latin America (LAC) continues to position itself as a rising global breadbasket. At present TFP growth rates, LAC will be able to meet 127% of regional food demand through productivity growth, an increase of 11 percentage points since 2014. The LAC region and particularly the southern cone nations of Argentina, Brazil, Paraguay, and Uruguay comprise the world's largest net exporting zone of agriculture products (Regúnaga 2013). These countries and others in Latin America have the potential to vastly increase their productivity to sustainably supply food and other agricultural goods for their own populations and to a growing world. Harmonizing trade rules and improving the trade capacity of low-income countries, coupled with improvements in supply chains and infrastructure, will foster timely and beneficial trade to close food and agriculture demand gaps.

In 2030, North America is projected to reliably supply safe, abundant food for the world, producing nearly as twice as much food to meet its own food demand. However, the potential for a new era of trade protectionism has sent a chill through agricultural producers who fear they will lose access to traditional trade partners or fail to access new markets at a time when prices are low and farmers are struggling. Investments in R&D would become more critical than ever under these circumstances (Box 2.5).

Box 2.5 Public Research Sparks Innovation and TFP

Due to agriculture's dependence on limited resources like water and land, it may be unique in its reliance on productivity and innovation to meet the rapidly growing demand of consumers by 2050 (Fuglie 2018). Agri-food innovation systems rely heavily on public agricultural research and development (R&D) and extension systems as well as regulatory frameworks that incentivize risk taking innovation and investment. Such agricultural R&D investments require long gestation periods of more than a decade to realize the full benefits that these investments generate. Over time, they pay large dividends, including higher profits for farmers, more abundant food supply at lower cost for consumers, and more opportunities and a higher quality of life in rural communities.

Filomena do Anjos is a senior lecturer and veterinarian at Eduardo Mondlane University, Mozambique. She is developing a more economical poultry feed, as more than 70% of rural families in Mozambique raise chickens. Photo credit: Carlos Litulo

Agricultural R&D along with extension programs are essential public goods and the principal drivers of total factor productivity (TFP) growth. Public sector R&D and extension programs deliver innovation and information to agricultural producers. They provide access to proven techniques such as conservation agriculture and animal care practices to improve the sustainability and resilience of their operations. While farmers innovate on their

(continued)

(continued)

farms, experimenting with practices that can boost their own production, individually they do not have the capacity to conduct long-term research and development activities.

Public R&D provides foundational results that the private sector can further develop to improve specific crops, livestock, machinery, or food manufacturing industries. R&D and extension services help producers control costs, reduce loss and waste, and become resilient to weather challenges and climate change while conserving natural resources.

Countries that build national agricultural research systems (NARS) capable of producing a steady stream of innovations suitable for local farming systems, such as Brazil's EMBRAPA, have generally achieved higher growth rates in agricultural productivity than countries that do not make these investments.

Future TFP growth in North America will be driven by innovations such as advanced crop and livestock breeding and data systems that monitor plant growth and animal health. However, public sector investments in the research and development (R&D) that drive agricultural innovation has slowed in the United States and in many high-income countries (Fig. 2.8).

For high-income countries, the growth rate in spending for public agricultural R&D averaged 4% annually between 1960 and 1990. Between 1990 and 2009, the growth rate declined to just 1.3% annually and *then began to contract between 2009 and 2013, declining on average 1.5% annually.*

Public R&D provides discoveries that are the foundation for further private sector innovation; lower public investments constrict the innovation pipeline. Private sector research investments, while significant, cannot make up the public R&D funding gap. Increased public sector R&D investments are needed to reinvigorate productivity growth. Additionally, as urbanization increases, so does competition for land and water resources. Continued farm consolidation will create some additional efficiencies, but land and water-use policies must balance the resource needs of agricultural producers with those of their urban customers.

2.6.2 *Growing Productivity While Protecting Against Risk*

Globally, productivity growth must continue to be a priority to sustainably meet the demand for food, feed, fiber, and biofuel. Yet productivity alone is insufficient to achieve economically, environmentally, and socially sustainable food and agriculture systems.

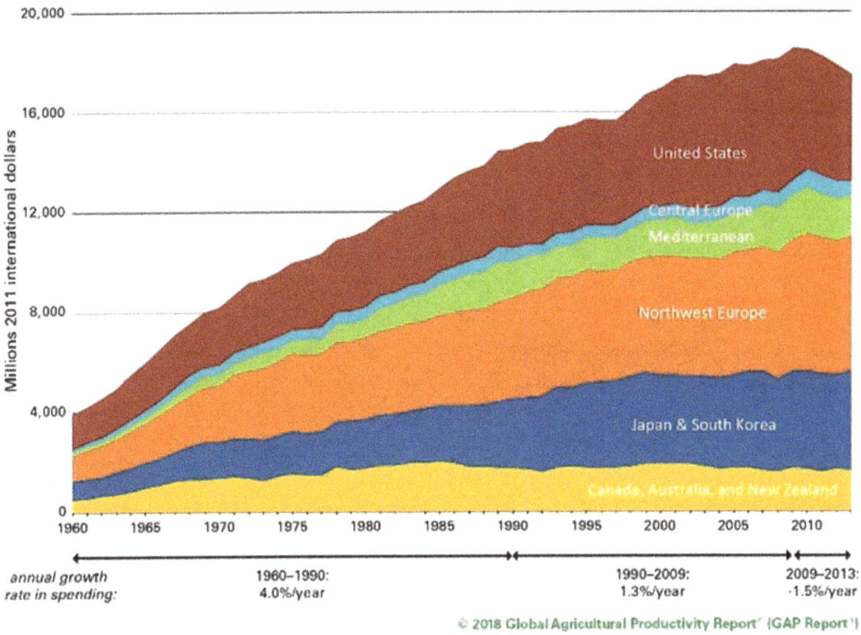

Fig. 2.8 Public agricultural R&D spending in high-income countries, 1960–2013. (Source: USDA ERS analysis of data from the Organization for Economic Co-operation and Development, Pardey and Roseboom (1989), World Bank and numerous supplementary sources. Paul Heisey and Keith O. Fuglie, "Agricultural research investment and policy reform in high-income countries." USDA ERS Research Report Number 249, May 2018)

Food and agriculture systems are vulnerable to a variety of risks, including extreme weather events and climate change, market volatility, and political instability. During times of crisis, agricultural producers seek to minimize their losses without putting their future productivity at risk. Good innovations and an enabling policy environment can ensure they stay productive during seasons of risk. This also helps stabilize the supply and price of food and agriculture products (Box 2.6).

Public and private insurance programs, such as crop insurance or weather index insurance, help preserve producer incomes and enable them to keep their most productive assets and to more effectively manage risk. Some producers participate in conservation programs that reward them for protecting their soil and water resources. Those without access to insurance and conservation programs face difficult choices. During hard times, small-scale farmers usually raise cash by selling cattle and equipment or by leasing their land; the poorest farmers have little to sell. Instead, they reduce their consumption of food and may resort to pulling children from school and into labor. They also reduce the already less than optimal proportion of family income spent on providing access to health services for children and women, particularly lactating and pregnant women. These coping strategies have negative, long-lasting impacts on the health and economic prospects of the family as well as their farm operations.

Consumers also face risks from economic instability or food price shocks. Governments are establishing social protection programs to stabilize households experiencing food and income insecurity. Some countries rely on national reserves to feed their population and manage food prices. Ensuring that agricultural trade remains open is essential to keeping food prices stable, especially when commodity stocks are low.

Box 2.6 Social Protection Programs Reduce Risk (Daidone et al. 2014)
Social protection programs, such as cash grants, provide the poorest rural residents with income stability and food security while also reducing their reliance on agricultural wage labor and freeing them up to invest time and resources in their own farms, to develop off-farm enterprises, or to pursue training for non-agricultural employment. In 2010, the *Zambian Ministry of Community Development for Mother and Child Health (MCDMCH)* piloted a *Child Grant Program (CGP)* in three provinces, where the program gave households with children under the age of 5 a total cash grant of $12 per month. Payments were made monthly and without condition.

The Child Grant Program, a pilot project in the Eastern Province of Zambia, provided families who have children under the age of 5 with a monthly grant that helped stabilize family incomes, enabling parents to invest more time and resources in developing their farms and off-farm enterprises.

The program not only reduced the severity of poverty, but it changed the participants' perception of their own food and income security: the number of households that reported being better off than they had been 12 months earlier increased by 45%. Perhaps most encouraging was the increased investment in productivity-enhancing and labor-saving inputs and the increases in agricultural output by CGP beneficiaries. The value of the overall harvest increased by 50%, on average, with most of the additional production being sold. CGP households increased both their ownership of livestock (21%) and the diversity of their livestock.

Finally, the income stability of the cash grants enabled participants, particularly women, to reduce their wage labor hours and develop their own

(continued)

(continued)

enterprises. The percentage of households that operated off-farm businesses increased by 17%. The CPG grants also had a significant multiplier effect: each Zambian kwacha transferred to a recipient generated 1.79 kwacha in the local economy.

In order to move people from "protection to productivity," social programs must be accompanied by investments and partnerships that improve producers' access to secure land tenure, transportation, electricity, and irrigation infrastructures and agricultural knowledge and innovations developed and disseminated by a robust research and extension system.

2.7 Food Wasted Is Productivity Lost

Reducing agricultural losses on the farm and food lost throughout the agricultural value chain avoids wasted resources and unnecessary greenhouse gas emissions.

On average, Americans throw away one pound of food each day, the equivalent of 30% of the calories they normally consume. Fruits and vegetables alone account for almost 40% of the waste, 17% is milk and dairy products, and 13.5% is meat (Conrad et al. 2018). These wasted foods are important sources of vitamins, minerals, protein, and calcium that promote healthy lifestyles and reduce healthcare costs. In addition, they represent a waste of all the water needed to produce them in the field and to process them in the value chains connecting farm produce with consumers.

Higher-income households tend to replace spoiled foods quickly, for example, purchasing another carton of strawberries during the next trip to the grocery store when the carton in the refrigerator goes bad. In those households, the lost nutrients are replaced and therefore more likely to be consumed. But the price of highly perishable foods can prevent many households from replacing spoiled food right away, so the opportunity to consume those nutrients is lost, along with the food.

Not only do spoiled foods end up in landfills producing methane, but they are a waste of the agricultural resources used to produce them.

A USDA study calculates that the equivalent of 30 million acres of cropland would be needed to produce the food and animal feed for livestock products (dairy, meat, and eggs) that Americans throw away each year (Conrad et al. 2018). Nearly 4.2 trillion gallons of irrigation water is wasted, including 2.3 trillion gallons to produce the wasted fruit and vegetables alone. Wasted fruits and vegetables are responsible for most of the pesticide waste, while most of the wasted cropland and fertilizer is used to produce feed for livestock.

While beyond the scope of this chapter, wasted food is also a waste of agricultural labor, capital (mechanization), and public and private sector investments in the development of technologies for agricultural productivity and sustainability (Fig. 2.9). The economic and environmental costs of transporting, packaging, and storing food that eventually ends up in the garbage and landfills also need to be taken into account in the cost of wasted food.

Improvements must be targeted throughout every part of the value chain: better harvesting and storage practices, better livestock care to reduce disease, improvements to the cold chain and the transportation infrastructure it relies on, reductions in waste at the processing and retail levels, and changes in consumer behavior.

Reducing loss and waste on a wide scale depends on government investments in public goods, such as infrastructure. An enabling policy environment that supports private sector innovation in harvest and storage technologies and stimulates behavior change by consumers is also vital. In addition, there are opportunities to increase the productive use of unconsumed food and agricultural byproducts (Box 2.7). These are potential sources of bio-energy, animal feed, fertilizer, and new products.

Reducing loss and waste and creating more opportunities to use waste productively will help meet the growing global demand for agricultural products, generate

Cropland	30 million acres
Irrigation Water	4.2 trillion gallons
Pesticide	780 million pounds
Fertilizer (Nitrogen, Phosphorus, Potash)	5.6 billion pounds

Fig. 2.9 Agricultural inputs and resources used to produce food waste in the United States, annual average, 2007–2014. (Source: Conrad et al. (2018))

Box 2.7 Cutting Food Loss Improves Nutrition Too!

In Nigeria, nearly 30% of children under the age of 5 are vitamin A deficient, a condition that can lead to blindness and increased risk of disease and premature death (Maziya-Dixon et al. 2006). Tomatoes are an excellent source of vitamin A, and Nigerian farmers produced 1.8 million metric tons of tomatoes in 2010, making their country the 16th largest producer in the world (Ugonna et al. 2015).

Photo credit: Global Alliance for Improved Nutrition (GAIN).

(continued)

(continued)

But the tomato supply chain is poorly organized and underdeveloped, and as a result half of the annual tomato harvest never reaches the market. Meanwhile, Nigeria imported 150,000 metric tons of processed tomato products in 2014, valued at $160 million (Ugonna et al. 2015).

The Geneva-based *Global Alliance for Improved Nutrition (GAIN)* has convened a coalition to develop solutions for reducing tomato losses that are market-based, nutritionally focused, locally adaptable, and financially sustainable. The *Postharvest Loss Alliance for Nutrition (PLAN)* brings leaders from government, finance, and academia together with representatives from Nigeria's tomato industry, including aggregators, processors, packagers, and cold chain operators.

The Alliance is targeting specific elements in the supply chain for improvement: crating and cooling technologies to protect prevent spoilage; a larger more reliable fleet of transport vehicles; new processing technologies and financing models to increase capacity; and outgrower schemes to link processors with farmers. Growers, traders, and processors also need technical assistance in negotiating contracts, tracking inventories, re-tooling and maintaining machinery, food safety protocols, and networking within the industry. Businesses with the capacity to scale up and innovate are receiving technical assistance and access to grants or affordable financing so that they can experiment with technologies and implement new approaches.

Strengthening the tomato value chain will not only give Nigerian producers access to a robust and growing market, but it will also provide low-income consumers a safe, affordable source of nutritious food that will improve the health of millions of children.

clean energy, mitigate carbon emission, create new jobs and industries, and improve incomes and food security, especially for small-scale producers.

2.8 Sustainable Agriculture Is Built on Productivity

Sustainable agriculture must satisfy human needs, enhance environmental quality and the natural resource base, sustain the economic vitality of food and agriculture systems, and improve the quality of life for farmers, ranchers, forest managers, fishers, agricultural workers, and society as a whole.[1] Improving agricultural sustainability requires multi-faceted, collaborative solutions involving producers, agribusinesses, transporters, retailers, and policymakers.

[1] Based on the definition in *Toward Sustainable Agriculture Systems in the twenty-first Century*, National Research Council, USA, 2010.

The United Nations defines sustainable growth as "meeting the needs of the present without compromising the ability of future generations to meet their own needs." No one understands this delicate balancing act better than farmers, ranchers, forest managers, and fishers.

As they balance the demands of the present with the needs of the future, producers decide how much risk they are willing to take. They must consider the risk management options available to them, as well as factors they cannot control like weather, market prices, and economic or political uncertainty. While trade-offs are inevitable, policies and investments that support agricultural productivity and expand risk management capacity give producers the best chance to meet current and future needs while increasing their adaptability and resilience.

The next section of the chapter outlines how farmers in Colombia, the United States, Kenya, and India are adopting innovation to build productive sustainable agriculture systems.

2.8.1 Making Colombia's Beef More Sustainable

With abundant natural resources and increased political and economic stability in many countries, the Latin America and the Caribbean region (LAC) looks to agriculture as a key opportunity to feed its expanding middle class and become a breadbasket to the world. Agricultural productivity growth on the continent has skyrocketed in recent decades, and the region is now beginning to shift toward lower-carbon, environmentally friendly agriculture systems (Truitt Nakata and Zeigler 2014).

Despite this progress, difficult issues must be addressed. Conserving forests and biodiversity while improving livestock productivity in Latin America will be key to cultivating a successful sustainable agriculture system. While Latin America produces more beef than any other region, emissions from beef production are the second highest in the world after South Asia. Nearly one-third of Latin America's beef sector emissions come from land-use change for pasture expansion (Gerber et al. 2013).

The problem is acute in Colombia, one of the world's top cattle-producing countries with 23 million head of beef and dairy cattle. Cattle production uses 28% of Colombia's total land area (Nelson and Durschinger 2015), with 80% of all agricultural land in Colombia used for pasture.[2] Decades of civil conflict have exacerbated forest and biodiversity loses, with some three million hectares (7.4 million acres) of forest destroyed.

Colombia faces a challenge in helping its small- and medium-scale farmers shift to sustainable lower-carbon cattle production systems that use less land, conserve more forests, and provide higher incomes.

In the recent decade, many Colombian ranchers have begun to work with local cattle trade associations and the national Colombian Cattle Ranching Association

[2] Government of Colombia, Census of Agriculture (2014).

(Federación Colombiano de Ganaderos, FEDEGAN) as well as a Colombia sustainable livestock foundation (La Fundación Centro para la Investigación en Sistemas Sostenibles de Producción Agropecuaria, CIPAV) to implement a more resilient form of livestock production: the silvopastoral production system (SPS) (Box 2.8).

Rows of fodder shrubs interspersed with grasses and trees characterize silvopastoral systems.

Photo credit: Neil Palmer, CIAT

With technical advice from CIPAV's SPS experts, ranchers plant fodder shrubs in high densities and intercropped with grasses and trees in rows. Using special fodder shrubs like *Leucaena leucocephala* and grasses like *Brachiaria* (high-protein fodder and grasses developed by CGIAR (Consultative Group for International Agricultural Research) institutions such as the International Center for Tropical Agriculture (CIAT)) boosts forage nutrition for cattle, allowing them to gain weight and produce more milk and meat in less time.

Leucaena shrubs grow rapidly and help fix nitrogen to soil, enriching soil health. Such forages grow deep root systems that help prevent soil erosion and can be integrated in other silvopastoral systems globally.

Box 2.8 Enabling Higher Productivity While Protecting the Environment in Colombia

Using a *healthy agricultural systems (HAS)* approach that focuses on increasing productivity while preserving the assets – the water, soil, and rich biodiversity that make productivity possible – *The Nature Conservancy (TNC)* and its partners are enabling farmers of all sizes to adopt practices that repair the land and sequester carbon, thereby ensuring more productive and profitable farm operations.

In Colombia, TNC and partners have supported 2600 ranchers in their transition to healthy agricultural systems over the past 7 years. Results have been impressive. Milk and meat production increased by 20%. Bird species numbers increased from 140 to 193, and the number of terrestrial mollusks,

(continued)

(continued)

ants, butterflies, and other wildlife increased. Monitoring studies have confirmed reduced pollution of water sources.

Healthy agricultural systems in Colombia optimize natural ecosystems to restore vitality to landscapes, increase productivity and farm profit, slow deforestation, and boost sustainability.

Photo credit: Ganadería Colombiana Sostenible

The climate impact of the healthy agricultural systems approach is equally impressive. To date, farmers have contributed to capture 1.5 million tons of CO_2 equivalent by converting degraded pastures into silvopastoral systems (grazing systems incorporating special fodder, grasses, and trees with rotational plots for livestock). They have avoided additional emissions by planting secondary forests and by preserving the natural forests within the project areas. Both contributions are highly significant for Colombia, as the country's climate change commitment for the cattle ranching sector is to mitigate 10.3 million tons of CO_2 equivalent by 2030.

The Nature Conservancy is working to expand these practices across Colombia and other countries in Latin America, demonstrating that agriculture and natural habitat can work hand in hand to preserve the planet while increasing production to feed a growing world.

Trees provide shade for the cattle, protecting them from heat. And with more vegetation in the pastures, the soil retains nutrients, water, and carbon, making ranches more resilient to cyclical drought.

The Government of Colombia has proposed reducing the total land used for livestock by 21% by 2030, and the national cattle ranching association, FEDEGÁN, proposes similar pastureland reduction goals along with productivity increases. But making the transition from extensive cattle ranching systems to the newer silvopastoral systems is not easy, as it requires both technical support and a change in mindset. Many ranchers perceive forestry and cattle ranching as incompatible practices and often clear forested areas so cattle can graze on grasslands. Ranchers also fear that by using less pasture and conserving more forests, they risk losing some of their farmland to the government or other ranchers.

Government policy can help ranchers shift to silvopastoral systems with less risk. In Colombia, the Ministry of Agriculture and Rural Development is implementing more opportunity for ranchers to formalize ownership of land through secure land titling and helping ranchers to gain greater access to finance, as well as to certify they did not gain land through deforestation. Pilot programs are now available that provide low-interest loans and technical assistance to ranchers who want to convert their operations to silvopastoral systems. Eventually, more retail chains may be incentivized to purchase zero-deforestation beef, similar to retail agreements in Colombia with coffee growers.

By focusing on reducing costs, by providing quality meat and milk, and by certifying zero-deforestation branded meat and milk, Colombia's ranchers may be able to compete with imported products for the rising number of middle-class consumers. Implementing silvopastoral systems is an example of how innovation and productivity benefit farmers, consumers, and the environment.

2.8.2 How Innovation Grows More Sustainable Pork in the United States

Decades of public research and development along with research and growing partnerships with the private sector have resulted in high levels of pork productivity in the United States. Today is only takes five breeding hogs to produce the same amount of pork from eight hogs in 1959, or 38% fewer breeding animals.[3] As recently as 1989, the United States was a net importer of pork; today it is a net exporter, reaching more than 100 countries.[4] Consumers in these markets trust the safety and quality of US pork products, and demand continues to grow.

This substantial increase in pork productivity demonstrates how TFP works and the economic and environmental benefits of productivity growth.

Widespread adoption of innovative technologies and practices has increased pork output using the same amount or less land, labor, fertilizer, feed, machinery, and livestock. Efficient uses of these inputs generated cost savings for producers and consumers and improvements in the environmental footprint of the pork and animal feed value chains.

Pork productivity begins in the genes. Genetic researchers and veterinarians analyzed hundreds of animal traits to select and mate pigs to breed descendants that are healthier, use less feed, and produce more meat. Heritage breeds are cross-bred to create the best meat flavor and quality for consumers.

The pork feed value chain has also experienced a dramatic increase in productivity and sustainability (Box 2.9). Over the past 30 years, productivity-enhancing crop technologies and practices reduced the amount of land, labor, machinery hours,

Box 2.9 Building Sustainability Through the Pork Value Chain in the United States

Private sector investment, innovation, and scale are helping more farmers and ranchers shift to lower-carbon production systems. Smithfield Foods, the world's largest pig producer and pork processor, led the protein industry as the first to announce an ambitious greenhouse gas (GHG) emission reduction goal throughout its entire supply chain (*The 25 by '25 Initiative*). By 2025, Smithfield will reduce its absolute GHG emissions from its 2010 baseline by 25%, or four million metric tons, equivalent to removing 900,000 cars from the road.

The initiative began with the creation of a robust model to estimate the GHG footprint of Smithfield's entire supply chain – a collaboration with the University of Minnesota's NorthStar Institute for Sustainable Enterprise and in partnership with the Environmental Defense Fund (EDF). To ensure Smithfield reaches this goal, the company launched Smithfield Renewables, a platform within the organization that will unify, lead, and accelerate its carbon reduction and renewable energy efforts.

Smithfield made commitments to improve the carbon footprint of the feed crops for their pork production, optimize fertilizer use and improve soil health, install efficient manure management technologies, and more efficiently track and manage logistics of transportation fleets to cuts costs and emissions.

In 2017, Smithfield fed its pigs more than 7.4 million pounds of grain. The GHG analysis of the Smithfield supply chain noted that animal feed accounts for 15–20% of their entire production carbon emissions. By helping farmers in their feed supply chain shift to efficient fertilizer and soil health practices (such as using cover crops, nitrogen sensors, and other conservation practices) and by promoting sustainable grains such as sorghum (a resilient crop that costs less to grow, offers good nutrition for pigs, and serves as part of a crop-diversification strategy), the program provides a triple win: more profit for farmers, improved soil and water health with less greenhouse gas emissions for the planet, and nutritious sources of feed for healthy pigs.

fuel, and fertilizer used to produce hog feed. Alfalfa, corn, and soybean seeds improved through biotechnology and conventional breeding become healthy crops that are pest-resistant and herbicide-tolerant. Best practices for fertilizer management ensure that the right amount of the appropriate fertilizer is used at the right time and in the right place.

Machinery equipped with precision systems, such as GPS, cover every inch of the field with precisely planted seeds and treat each plant with the nutrients and crop protection products needed. Precision systems also allow less productive land to be identified and set aside for conservation use, such as pollinator or wildlife habitat.

These crops are blended with nutrients to make pig feed that is healthier and easier to digest, resulting in fewer methane emissions during the digestive process. "Smart barns" provide consistent temperature, comfortable housing, and readily available feed and water. With detailed data on the health and development of the herd, farmers can reduce energy use, save labor, and protect pigs from disease.

2.8.3 *Investing in Productivity for Africa's Dairy Hub: Kenya*

Kenya's dairy farmers produce more than five billion tons of milk per year, the most in Africa (FAOSTAT 2017). The dairy industry accounts for 6 to 8% of Kenya's GDP and provides income for two million households. Consumers also benefit from Kenya's dairy productivity; per capita milk consumption is 100 liters (26 gallons) per year, more than any other developing country (Katothya 2017).

Kenya's dairy industry is endangered by climate change. A substantial increase in mean temperature is predicted for East Africa and could lead to a reduction in fodder output and grazing land capacity. Increasing temperatures threaten the health and productivity of livestock. As droughts lengthen and intensify, large-scale cattle losses are likely. Small-scale farmers will be forced to sell cows or land to cope with the loss of income, making it difficult for them to recover financially when the drought is over.

As part of its climate change adaption and mitigation strategy, Kenya's dairy sector needs to increase the productivity of its dairy cattle and reduce the GHG emission intensity of milk production (Box 2.10). Sub-Saharan Africa's milk production has the highest emission intensity in the world, three times greater than the global average and almost double that of South Asia.

Kenya is home to 75% of the dairy cattle in Southern and Eastern Africa; 80% of Kenya's milk output is produced by small-scale farmers. By improving cattle productivity and reducing emission intensity, the dairy sector in Kenya can significantly mitigate greenhouse gases while increasing small-scale farmer income.

More than half of the emission intensity of milk production in sub-Saharan Africa comes from methane produced during a cow's digestive process. One strategy for reducing these emissions is to add legume silages to a cow's diet. Legumes are digested more efficiently, so a cow produces less methane and more milk. Improving the genetics of dairy cattle is another way to reduce methane emissions

Box 2.10 Better Breeds and Better Feed Are Key for Climate Resilience

Photo credit: International Livestock Research Institute (ILRI)

The drylands of northeastern Kenya are particularly vulnerable to climate change. This region receives less than 500 millimeters (20 inches) of rain per year and has fewer than 90 plant growth days. Many of the people in this region are pastoralists, moving regularly to find forage for their livestock.

Boran cattle are well-suited to the dryland areas of East Africa but produce very little milk and meat compared to cross-bred and high-grade varieties. The International Livestock Research Institute (ILRI) farm in Nairobi is breeding Boran cattle that efficiently digest the low-quality grasses and silages that are common to the drylands. This will decrease methane emissions and improve milk and meat productivity.

In addition to improved cattle genetics, improving the fodder and feed for cattle is key to achieving more robust milk products and livestock that are climate-resilient. The International Center for Tropical Agriculture (CIAT) has developed *Brachiaria* grass varieties that are drought-resistant and increase milk productivity in dairy cows by 40%.

Photo credit: Georgina Smith/CIAT

(continued)

(continued)

These nutritious grasses are easier for cattle to digest. Demand for these improved grasses is skyrocketing, and farmers are now diversifying their income by growing and selling them at the local "fodder stores" which purchase and sell fodder in local markets.

and increase milk productivity. Kenya has already made strides in this direction; the country is home to more than 70% of the cross-bred and high-grade dairy cows in Africa. Sixty percent of the milk produced in Kenya (three billion liters) comes from high-grade cattle and cross-breeds. But high-grade dairy cattle are more susceptible to disease than local cattle varieties, so breeding for disease resistance is a top priority.

The health of cattle in Kenya and exposure to disease continue to threaten productivity and farmer livelihoods. Part of the county is infested with tsetse fly, the biological vector of trypanosomiasis (sometimes called sleeping sickness), a parasitical disease that causes anemia and emaciation in cattle. The condition is chronic, and if left untreated, it is often fatal. If a cow survives the infection, its milk productivity can drop by 30 to 40%. Trypanosomiasis is a zoonotic disease that is passed between animals and humans via the tsetse fly, although the number of human cases in Africa has dropped substantially due to sustained public health efforts.

To ensure a sustainable livelihood and earn sufficient incomes to invest for the future, small-scale dairy farmers need consistently healthy, productive herds. Good animal care and feeding practices promote productivity and prevent disease, but access to affordable, quality animal healthcare products is also essential.

2.8.4 *Mechanizing for the Future in India*

India's small-scale farms have enjoyed healthy yields, thanks to the Green Revolution and continued improvements in seeds, crop protection products, and access to fertilizers (Box 2.11). Nevertheless, labor productivity on small farms remains stubbornly low. Family members do the bulk of the farm work because mechanization rental and ownership are more expensive than family or hired labor.

Not only is this an inefficient use of labor, but it contributes to high rates of rural poverty and food insecurity. For example, the income from a one-hectare farm, even if it is high-yielding, must meet the needs of as many as 12 people. As a result, small farmers are heavily dependent on food rations, wage labor, and government support to supplement their farming incomes.

Custom Hiring Centers (CHCs) give farmers affordable access to mechanization without having to own the machines themselves. Farmers can rent tractors and implements for soil preparation, seeding, application of nutrients, and crop protection and harvesting. CHCs are centrally located to serve several villages, reducing

Box 2.11 Balanced Crop Nutrition Boosts Productivity and Incomes in India

Since 2008, the *Mosaic Villages Project*, a collaboration between The Mosaic Company, The *Mosaic Company Foundation*, and implementing partner, the *S M Sehgal Foundation*, has helped Indian farmers move out of poverty and achieve greater food security. Mosaic's investment includes funding and the expertise of Mosaic agronomists who work alongside local partners to train farmers in balanced crop nutrition and agronomic best practices.

Photo credit: The Mosaic Company Foundation

In the remote districts of Mewat and Alwar in Rajasthan, the *Krishi Jyoti Project*, or "enlightened agriculture," helps farmers improve productivity of three crops: pearl millet, wheat, and mustard. The project focuses on five key aspects of agricultural production: soil health, seed and fertilizer, water management, agronomic training, and market linkages. Village leaders selected farmers representing all castes and landholding sizes to participate in the program.

With balanced crop nutrition practices – using the right mix of macro- and micro-nutrients to meet the needs of the crops and soils – together with agronomic expertise and financial support, farmers increased yields by as much as 25% over traditional farming practices. In total, Krishi Jyoti has directly benefited more than 26,000 farmers across 60 villages and boosted cultivation across nearly 16,000 acres of land. Average income per acre has also grown between 4480 Rs ($70 US) for wheat to 5760 Rs ($90 US) for mustard.

Communities participating in the Krishi Jyoti Project are using the additional income to help create a better life for future generations. The S M Sehgal Foundation and Mosaic funded renovations for 20 schools in Alwar, Mewat, and Sonipat – including adding sanitation facilities, safe drinking water systems, and school kitchens.

the time and cost of transporting the equipment. CHC partnerships include equipment manufacturers, such as *John Deere*, who provide the equipment, product service, and training in agronomy practices and equipment usage. State governments contribute financial support and invest in infrastructure for the centers and in road improvements to ensure that equipment can be transported efficiently. Local entrepreneurs are hired to operate the centers, deploy and maintain the equipment, as well as manage the contracts with the farmers.

Nearly 90% of farmers with less than two hectares participate in a government food ration program. India has 120 million individual landholdings under two hectares. To meet its targets for reducing food insecurity and poverty, the government needs to invest in non-agricultural employment and skills training for rural workers to move more people out of agriculture, particularly manual labor, while fostering off-farm agricultural employment in jobs such as agro-dealerships, equipment and machinery maintenance, processing, and storage.

Farmland consolidation can help achieve greater economies of scale as well. The necessity for consolidation is amplified by the growing competition for land. India's rapidly expanding manufacturing and service industries need room to grow and are already competing for land and displacing farmers across the country.

2.9 Policies to Create an Enabling Policy Environment for Productivity in Agriculture

Meeting the rising demand for nutritious, affordable food as well as materials for fuel, clothing, housing, and consumer products will require innovative, productive, and sustainable food and agriculture systems. Together, governments, producers, and the entire agri-food system must commit to improve and transform the system to achieve a healthy population and a healthy planet. Improving agricultural sustainability requires multi-faceted solutions built on science-based public policies.

Productivity in agriculture grows when governments invest in public research, development, and extension services; when all participants in the agri-food system embrace, customize, and disseminate science-based and information technologies; when the private sector can be incentivized to form partnerships for infrastructure development and improved nutrition; and when capacity for regional and global trade in agriculture is streamlined.

2.9.1 Smart Regulatory Systems Build Trust and Competitiveness for Productivity

Governments establish agricultural policies and regulations to ensure human health and safety, protect the environment and animal welfare, and foster economic growth while meeting consumer needs for food, fiber, fuel, and other coproducts. Smart regulatory systems that keep pace with rapidly changing innovations in science and technology can foster the adoption of such innovations (Box 2.12).

Box 2.12 A Twenty-First-Century Regulatory System for Agriculture

For thousands of years, agriculturalist have improved the quality and performance of crops and livestock through trial and error, saving seeds from plants or breeding animals from those that exhibited the desired traits. Today, the tools used by agricultural breeders have evolved through science-based innovations. With an ability to understand the genetic sequence of plants and to link a particular gene with a specific plant characteristic, breeders can quickly and efficiently improve plants while avoiding the transfer of unwanted genes.

In the past decade, *new gene-editing techniques such as CRISPR-Cas* (clustered regularly interspaced short palindromic repeats, DNA sequences that can be used to instruct genes to perform beneficial functions and more precisely edit DNA) have become available, unlocking potential benefits for farmers, consumers, and the environment. Breeders can now edit genes by turning on or off various genetic functions that increase crop yields during drought, protect the plant or crop against viruses and pests (reducing the amount of pesticide needed), improve the nutritional quality and content of crops, or help vegetables maintain longer shelf life. Gene-editing technologies such as CRISPR-Cas rely on natural processes that happen in the genome but channel and target those changes more precisely.

Seed companies are exploring how this technology allows breeders to develop better hybrids by quickly finding and leveraging the inherent diversity existing in crops. Livestock breeders are also harnessing the power of gene-editing tools to improve the resilience, productivity, and nutritional content of animals for better meat, milk, and eggs.

To fight a devastating corn disease affecting small-scale farmers in Africa (maize lethal necrosis), Corteva Agrisciences™ and the *International Maize and Wheat Improvement Center (CIMMYT)* have formed a public-private research partnership to improve the resilience of maize to this devastating disease using CRISPR-Cas technology.

(continued)

(continued)

Plants and animals derived from new breeding methods such as biotechnology and gene-editing should be assessed for potential health or safety impacts, rather than for the processes used to produce the trait or product. Without streamlined modern regulatory systems, innovation from small companies, public-private research partnerships and universities many not reach farmers who need solutions. In the United States, the regulatory system for biotechnology had not been revised since 1986 and required modernization to address new breeding technology. In 2020, the USDA finalized new regulations under the Sustainable, Ecological, Consistent, Uniform, Responsible, Efficient (SECURE) rule, providing new guidance for plant breeding innovation.

Plant breeding innovations like CRISPR-Cas will only be achieved through improving and updating regulatory systems and active engagement and collaboration with farmers, academia, governments, NGOs, and public research institutes, both in the United States and around the world.

A successful regulatory system establishes predictable, clear, science-based operating conditions for farmers and ranchers – particularly with regard to seeds, crop protection, and animal health – as well as for mechanization companies, insurance and finance firms, and food processing and retail industries, so that the overall agriculture sector can deliver value for people, the environment, and the national economy.

In today's global competitive environment, regulatory systems are being called upon to do even more, as consumers seek more information about production methods, nutritional content, labor practices, and sustainability of local, national, and international food and agriculture systems. Transparency and traceability are growing in importance for developing consumer trust, while affordability and accessibility remains a paramount concern for many customers.

It is especially important that government regulatory systems help foster productivity and innovation while avoiding unnecessary costs, delays, and burdens to the agriculture sector, ultimately impacting the ability to swiftly deliver quality products to consumers. Regulatory systems should have a sound legal and empirical basis, minimize costs and market distortions, and promote innovation through intellectual property protection and market incentives. They must be clear and practical for users and be compatible with domestic and international trade principles.[5] Smart regulatory systems contribute to innovation and productivity when all the participants – government, industry, producers, scientific researchers, members of the media, and consumers – responsibly engage in practice as well as understanding about new opportunities that science and technology bring (Fig. 2.10).

Ideally, farmers practice good stewardship with innovation technology; input providers, processors, and retailers work within regulatory frameworks; government consults with all relevant parties and establishes well-functioning, science-

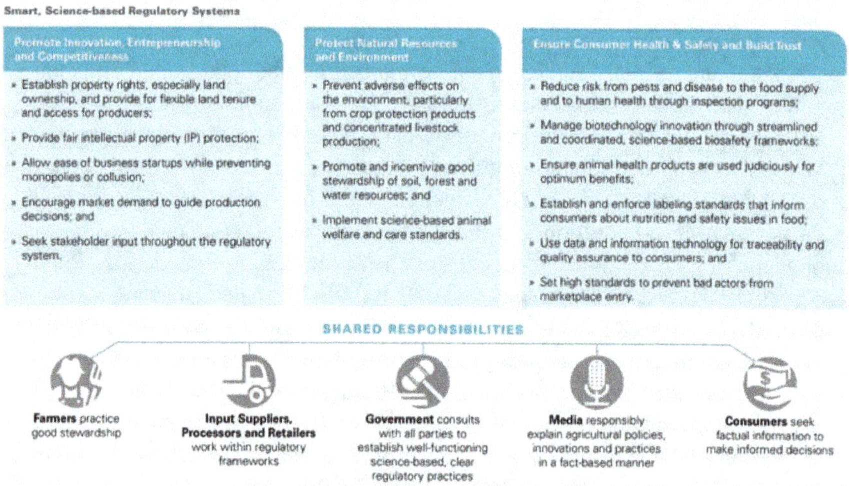

Fig. 2.10 Science-based regulatory systems for productivity in agriculture. (Source: 2016 Global Agricultural Productivity Report (GAP Report))

based, and clear regulatory practices; media responsibly explain agricultural policies, innovations, and practices in a fact-based manner; and consumers have easy access to facts to make informed decisions.

2.10 Investing in Farmers

To maximize the productive potential of investments in agricultural R&D, innovations such as mechanization and improved practices for soil and animal health, governments, and the private sector need to invest in the productivity of farmers, ranchers, foresters, and fishers.

Farmers and producers are already the largest investors in the agricultural value chain. A review of agricultural investment sources in low- and middle-income countries by the *UN Food and Agriculture Organization (FAO)* found that 78% of agriculture investments come from on-farm investment in agricultural capital by farmers themselves (Lowder et al. 2012) The remaining 22% comes from government expenditures, public sector agriculture R&D, foreign direct investment, and official development assistance.

Yet, millions of small-scale farmers, especially women and young people, are undercapitalized because they do not have legal title to their most important capital asset, their land.

In low-income countries, the right to occupy, cultivate, inherit, lease, buy, or sell land is often determined by a complex system of social customs that are granted and arbitrated by communal authorities (sometimes called "customary" or "tribal" authorities). Communal rights may be recognized by civil authorities as well, but they do not have the same legal standing as land titles or leaseholds granted by the state (Cousins 2016). Formal lenders see communal tenure rights as a risky investment and are reluctant to extend credit, regardless of the productive potential of the land.

Communal tenure systems are often dominated by social and patriarchal hierarchies that disenfranchise vulnerable groups. As a result, gender, age, and community standing often determine the quality, quantity, and terms of the landholdings.

Furthermore, without civil legal protections, communal landholders have little recourse if their land is appropriated by customary or state authorities. In these circumstances, communal landholders, particularly women, are less likely to make investments in improved seeds or fertilizer, suppressing their earning potential and making it difficult to save for capital purchases, such as mechanization and irrigation technologies. Likewise, long-standing and widespread land disputes stifle the sale, purchase, leasing, and inheritance of land, effectively freezing the land market, discouraging productive investment, and stifling economic growth (Valetta 2012).

Policies that secure and promote farmer access to land, water, and improved inputs enable farmers of all scales to remain competitive even during the challenging phases of business cycles and help them respond to changing climate patterns.

2.11 Concluding Remarks

Without productivity growth, the world's agriculture and food systems will not be able to sustainably produce the food, feed, fiber, and biofuel needed for ten billion people in 2050. An enabling policy environment also supports the productive potential of farmers, ranchers, foresters, and fishers by generating new market opportunities, increasing their access to affordable financing, and improving the environmental sustainability of their operations. The data show that a global focus on productivity growth is urgently needed to reverse the downward trend in productivity growth.

While this chapter has focused on the role of the producer, the role of the consumer in productive sustainable food systems is just as critical. Significant reductions in food waste are critical to sustainable food systems, and most food waste occurs at the retail and consumer point of the value chain. Consumers also need to educate themselves on the science and practice of agriculture and be willing to challenge their assumptions about what makes their food "sustainable." Without consumer acceptance of science-based innovations such as biotechnology, gene-editing, and proven practices in livestock husbandry, the target levels for agricultural productivity and sustainability will not be realized.

Innovation and productivity are interdependent, but access to and acceptance of innovative technologies and practices must be improved if they are to realize their potential.

References

Conrad Z, Niles MT, Neher DA, Roy ED, Tichenor NE, Jahns L (2018) Relationship between food waste, diet quality, and environmental sustainability. PLoS One 13(4):e0195405. https://doi.org/10.1371/journal.pone.0195405

Cousins B (2016) Land reform in South Africa is sinking. Can it be saved? Thought piece commissioned by Nelson Mandela Foundation, Council for the Advancement of the South African Constitution, and Hans Seidel Foundation, https://www.nelsonmandela.org/uploads/files/Land__law_and_leadership_-_paper_2.pdf

Daidone S, Davis B, Dewbre J et al. (2014) Zambia's Child Grant Programme: 24-month impact report on productive activities and labour allocation, FAO. http://www.fao.org/3/a-i3692e.pdf

FAOSTAT on line (2014). http://www.fao.org/faostat/en/#home

FAOSTAT on line (2017). http://www.fao.org/faostat/en/#home

Fuglie K (2018) R&D capital, R&D spillovers, and productivity growth in world agriculture. Appl Econ Perspect Policy 40(3):421–444

Gerber PJ, Steinfeld H, Henderson B, Mottet A, Opio C, Dijkman J, Falcucci A, Tempio G (2013) Tackling climate change through livestock – A global assessment of emissions and mitigation opportunities. Food and Agriculture Organization of the United Nations (FAO), Rome

Hunter CM, Smith RG, Schipanski ME, Atwood LW, Mortensen DA (2017) Agriculture in 2050: recalibrating targets for sustainable intensification. Bioscience 67(4):386–391

Katothya G (2017) Gender assessment of dairy value chains: evidence from Kenya, Rome, FAO

Landes M et al. 2017. India's dairy sector: structure, performance and prospects, USDA Economic Research Service (USDA ERS), LDPM-272-01

Lowder SK, Carisma B, Skoet J (2012) Who invests in agriculture and how much. FAO, EAS Working Paper No. 12-09

Maziya-Dixon B, Akinyele IO, Sanusi RA, Oguntona TE, Nokoe SK, Harris EW (2006) Vitamin A deficiency is prevalent in children less than 5 y of age in Nigeria. J Nutr 136(8):2255–2261. https://doi.org/10.1093/jn/136.8.2255

Naresh RK et al (2017) Climate change and challenges of water and food security for smallholder farmers of Uttar Pradesh and mitigation through carbon sequestration in agricultural lands: an overview. Int J Chem Stud 5(2):221–236

Nelson N, Durschinger L (2015) Supporting zero-deforestation cattle in Colombia. USAID-supported Forest Carbon, Markets and Communities Program. Washington

OECD/FAO (2016) OECD-FAO agricultural outlook 2017–2026. OECD Publishing, Paris. https://doi.org/10.1787/agr_outlook-2016-en

OECD/FAO (2017) OECD-FAO agricultural outlook 2017–2026. OECD Publishing, Paris. https://doi.org/10.1787/agr_outlook-2017-en

Regúnaga M (ed) (2013) Global food security and agricultural natural resources: role and views of Argentina, Brazil, Paraguay and Uruguay, 1a edn. Editorial De Yeug, Argentina, 160 p

Steensland A, Zeigler M (2018) Global agricultural productivity report® (GAP Report®). Global Harvest Initiative, Washington, D.C.

Truitt Nakata G, Zeigler M (2014) The next global breadbasket: how Latin America can feed the world. Inter-American Development Bank and the Global Harvest Initiative. https://publications.iadb.org/en/publication/17428/next-global-breadbasket-how-latin-america-can-feed-world-call-action-addressing

Ugonna C, Jolaoso MA, Onwualu PA (2015) Tomato value chain in Nigeria: issues, challenges and strategies. J Sci Res Rep 7(7):501–515. https://doi.org/10.9734/JSRR/2015/16921

Valetta W (2012) Rural land tenure security and food supply in southern Benin. In: Knowledge and innovation network, Volume II, Issue I, Spring/Winter 2012–2013.

von Lampe M et al (2014) Why do global long-term scenarios for agriculture differ? An overview of the AgMIP Global Economic Model Intercomparison. Agric Econ 45(1):3–20

Zeigler M, Steensland A (2017) Global agricultural productivity report® (GAP Report®). Global Harvest initiative, Washington, D.C.

3

The Role of Value Creation and Innovations in Crop Breeding

Mathias L. Müller and Hugo Campos

3.1 Origins of Agriculture

The evolution of human society is closely associated with, and has been dependent on, the progress man has made in agriculture. Therefore, it is no surprise that the field of agriculture has been a particularly fertile ground for innovation. The Neolithic Revolution, at the start of the Holocene calendar, about 12,000 years ago, marks a transition period during which our ancestors' lifestyles changed from hunting and gathering to cultivating crops and growing livestock (Bocquet-Appel 2011). The fertile crescent in the Middle-East, from upper Egypt to the Persian Gulf, including the southeastern regions of Turkey and the western fringes of Iran, are often called the "cradle of civilization" for the fact that sedentism and farming emerged there first (Maisels, 1993), fundamentally transforming the way of life of previously nomadic groups of hunter-gatherers. The dawn of agriculture (from Latin ager = "field" and cultura = "growing, cultivation") was made possible by a warming of the earth's temperatures by approximately 20 °C following the last ice age, as could be inferred from ice cores collected in Greenland (Platt et al. 2017). The increasingly stable climate made the cultivation of crops and the raising of animals feasible and more reliable. Over the next two millennia, similar developments to those in the fertile crescent occurred and gave rise to the eight centers of origin of agriculture and their derived crops (Vavilov 1951). While the transformation of lifestyles had a profound effect on the environment through the practice of

Mathias Müller is a Corteva Agriscience employee. Corteva Agriscience is a for profit agricultural company.

M. L. Müller (✉)
Corteva Agriscience, Johnston, IA, USA
e-mail: mathias.muller@corteva.com

H. Campos
International Potato Center, Lima, Peru

local deforestation and irrigation, it also led to fundamental changes in society and the way humans segregated tasks for greater efficiency. Though successful innovation examples can be found in every single aspect of agriculture, we will only describe some of those related to crop breeding, since a comprehensive discussion of past innovations in agriculture is beyond the scope of this chapter.

3.2 Innovations in Crop Breeding

Wild grains were collected and eaten from at least 20,000 BC by hunter-gatherer populations. From around 9,500 BC, the eight Neolithic founder crops, emmer and einkorn wheat, barley, flax, peas, chickpeas, lentils, and bitter vetch, were cultivated in the Eastern Mediterranean regions extending from current Greece to Libya. Rice was domesticated in China around 6,000 BC, followed by mung, soy, and azuki beans (Diamond 1997). Plants which initially succeeded were self-pollinators which were easy to grow, and their early domestication selected for properties such as shatter resistance in wheat for easy harvesting. One of the many benefits of agriculture is greater efficiency. Hunter-gatherer populations were regulated by the amount of food and the number of kills a tribe could achieve. For example, a tribe of 50 might have needed a territory of 100 square kilometers to survive (Tallavaaraa et al. 2018). This in turn meant that the earth could, at most, support a population of 20 million hunter-gatherers. Efficiency became key to population growth and success of the human race; innovations in agriculture made it possible.

It is noteworthy that modern agriculture is capable of supporting today's world population of 7.7 billion, but that important innovations and significant productivity increases are urgently needed to support a world population, projected by the United Nations Department of Economic and Social Affairs to be 9.7 billion in 2050.[1]

Progress in early agricultural practices was slow, progressive, and localized. Earliest cultivators likely combined hunting-gathering with what they managed to grow (Kavanagh et al. 2018), and it took thousands of years to progressively select seeds that were easier to cultivate and provided more nutritious food. Our ancestors did not live in a global society; hence the same innovations had to be made independently in different parts of the world, sometimes thousands of years and miles apart. Weiss et al. (2006) divide domestication of plants into a "gathering" stage, a "cultivation" stage during which plants were sown and harvested, and a "domestication" phase during which mutants with desirable characteristics were selected. For most major food sources, the domestication phase resulted in progressively better cultivars. The process of selection had begun, but it was slow and empirical. The first documented hybrid was achieved by Thomas Fairchild, a "gardener in London" who founded City Gardens, a commercial venture in which he planted and bred fruits and flowers. He collaborated with botanists and plant collectors through which he was able to obtain plants from around the world. In 1717, he successfully crossed a carnation pink (*Dianthus caryophyllus*) with a Sweet William (*Dianthus barbatus*), and the resulting plant, described as *Dianthus caryophyllus barbatus*, became known as Fairchild's Mule. Significant revolutions in breeding took place in the nineteenth century, starting with Austrian monk Gregor Mendel from Brno (today a historic town in the Czech Republic) who, between 1856 and 1863, formulated what is known as the three principles of Mendel's laws of inheritance, on which modern genetics is based. Mendel focused on seven clearly identifiable traits of pea plants: seed shape, flower color, seed coat tint, pod shape, pod color, flower location, and plant height to elaborate the law of segregation, the law of independent assortment, and the law of dominance (Mendel 1866) which form the basis of Mendelian inheritance patterns, where a trait is controlled by a single locus in an inheritance pattern. Mendel postulated the existence of discrete "factors" (today known as genes) that are passed along from generation to generation. William Bateson, an English professor of biology at Cambridge University and the first director of the John Innes Centre, a leading plant and microbial genetics research organization based in Norwich, United Kingdom, coined the word genetics in 1905.

3.2.1 Modern Breeding Approaches

Mendel's laws of inheritance remain the solid foundation upon which modern breeding techniques were developed. For many years, breeding was based on the selection and crossing of the best individuals of a population. The advent of hybrid crops, spearheaded by the development of hybrid maize in the early twentieth century, started a revolution which resulted in a significant acceleration of the incremental yield gain curve in those crops which where amenable to the new breeding techniques. The "Green Revolution" emerged in the 1950s and 1960s, from collaborations and the transfer of technologies between nations, scientists, and farmers. Norman Borlaug, the "Father of the Green Revolution," tirelessly promoted the development of high-yielding grain cultivars and the adoption of improved management techniques, including the use of fertilizers and pesticides and the deployment of irrigation. The selection of cultivars with greater requirements for nitrogen, but also vastly increased yield potential, is credited for saving over a billion people from starvation. The Green Revolution started in Mexico, in collaboration with the US Government, the United Nations, the Food and Agriculture Organization ("FAO"), and the Rockefeller Foundation. From there, the breeding and management techniques expanded to other Latin American countries, Asia, and Africa. One of the great successes of the Green Revolution is rice cultivar IR8, which was derived, in the Philippines, from a cross between the Indonesian variety Peta and the Chinese variety Dee-geo-woo-gen by scientists at the International Rice Research Institute ("IRRI"). While IR8, which is a semidwarf variety resistant to lodging, required the supply of fertilizer for optimal production, its harvested yield was almost ten times that of traditional rice.

After the Green Revolution, a large number of innovations dramatically enhanced the ability of plant breeders to accelerate breeding processes. Scientists used information encoded in DNA to capitalize on the genetic diversity contained within cultivars and to enhance the efficiency and effectiveness of plant breeding endeavors. These innovations helped reduce the number of breeding cycles required to bring to market a new variety and accelerated the development of novel elite parents and improved cultivars, fulfilling the needs of humankind. The combined set of modern breeding innovations enabled tremendous increases in genetic gain and resulted in the development of maize hybrids which exhibit commercially relevant levels of drought tolerance and contain stacked transgenic events conferring resistance to insects and tolerance to herbicides, mediated by multiple modes of action. Table 3.1 describes some of the pivotal innovations which have enabled the development of current breeding programs in crop species.

Table 3.1 Main innovations enabling current public and private breeding programs

Innovation	Impact in plant breeding	Key reference/ review article
Doubled haploid plants	Reducing breeding cycles and facilitating the expression of recessive traits	Kasha and Kao (1970)
Transgenic plants	Expanding the range of genetic variation available to plant breeders and farmers	Horsch et al. (1985)
Restriction fragment length polymorphisms	First truly molecular marker available to gain insight on the organization of crop genomes	Helentjaris et al. (1986)
Linear mixed statistical models	Improved statistical analyses, increasing the genetic signal to environmental noise ratio and the quality of phenotypic datasets	Gilmour et al. (1997)
Quantitative trait loci	Examining commercially relevant quantitative traits with molecular markers, establishing the basis of molecular breeding	Paterson et al. (1998)
Genetic modelling	*Ex ante* quantitative simulation of genetic models, genotype-to-phenotype relationships, and breeding scenarios	Podlich and Cooper (1998)
Sequencing of the first plant genome, *Arabidopsis thaliana*	Establishing the basis to understand the molecular basis of plant variation	The Arabidopsis Genome Initiative (2000)
Genomic selection	Using genome-wide, instead of discrete molecular markers information, in molecular breeding	Bernardo and Yu (2007)
Gene editing	Unprecedented precision to create and manage genetic variation and improve traits in crops	Wolter et al. (2019), Chen et al. (2019)

3.3 Collaborations

3.3.1 *International Organizations*

Innovations derived from collaborations have shaped the agricultural landscape since the dawn of times. While passionate scientists, in academia or in the private sector, often collaborated to advance their research (see Thomas Fairchild, who documented the first hybrid crops), so did seed companies and equipment manufacturers. In some instances, collaborations between individuals allowed the most significant progress; in others it was collaboration between institutions. Serendipity often played a role. For example, consider the remarkable story of John Washington Carver, who was born into slavery and became one of the most influential agricultural scientists of his time. It is certainly his friendship with young Henry A. Wallace that influenced the latter to experiment with cultivars of maize and later to bring hybrid maize lines to Midwest farmers through the Hi-Bred Corn Company,[2] which he founded. Upon visiting Mexico as the newly minted US vice president under Franklin Roosevelt's administration, Henry A. Wallace discovered that, while maize was an essential staple in Mexico, the yields obtained by local farmers were far below those achieved in the United States. This recognition and his passion for increasing the yields of crops to feed the world encouraged Wallace, with support of the Rockefeller Foundation, to help set up an experimental station in Mexico, which would encourage collaboration between local farmers and US scientists. Norman Borlaug was one of the first scientists to join the station as a plant pathologist, together with soil scientist William E. Colwell and potato breeder John Niederhauser. The three are largely credited for being at the origin of the green revolution for which Norman Borlaug eventually earned his Nobel Peace Prize in 1970. In the middle of the twentieth century, rapid population growth triggered fears of widespread global famine. The Mexican government had established the Office of Special Studies (OSS) to coordinate the program initiated by the Rockefeller Foundation. This in turn laid the groundworks for the establishment and support of the International Rice Research Institute (IRRI) in 1960 and the International Maize and Wheat Improvement Center (CIMMYT) in 1963. In 1970, the Rockefeller Foundation proposed a broader coalition to support an international network of agricultural research centers and, with the help of the World Bank, the FAO, and UNDP, established the Consultative Group for International Agricultural Research (CGIAR) to reduce poverty and achieve food security in developing countries. Today, the CGIAR comprises 15 independent nonprofit research centers (Table 3.2), conducting innovative research and collaborating on a global scale.

[1] Which later became Pioneer Hi-Bred International, Inc., and is now part of Corteva Agriscience.

Table 3.2 Global network of CGIAR centers

CGIAR centers	Headquarters location
Africa Rice[a] (West Africa Rice Development Association, WARDA)	Abidjan, Côte d'Ivoire
Bioversity International	Rome, Italy
Center for International Forestry Research (CIFOR)[b]	Bogor, Indonesia
International Center for Tropical Agriculture (CIAT)[c]	Cali, Colombia
International Center for Agricultural Research in the Dry Areas (ICARDA)	Beirut, Lebanon
International Crops Research Institute for the Semi-Arid Tropics (ICRISAT)	Hyderabad, India
International Food Policy Research Institute (IFPRI)	Washington, D.C., United States
International Institute of Tropical Agriculture (IITA)	Ibadan, Nigeria
International Livestock Research Institute (ILRI)	Nairobi, Kenya
International Maize and Wheat Improvement Center (CIMMYT)	Texcoco, Mexico
International Potato Center (CIP)	Lima, Peru
International Rice Research Institute (IRRI)	Los Baños, Philippines
International Water Management Institute (IWMI)	Battaramulla, Sri Lanka
World Agroforestry Centre (International Centre for Research in Agroforestry, ICRAF)	Nairobi, Kenya
World Fish Center (International Center for Living Aquatic Resources Management, ICLARM)	Penang, Malaysia

[a]AfricaRice and IRRI joined efforts in 2018 to offer a pan-African, multi-focus rice program
[b]On January 1, 2019, the Center for International Forestry Research (CIFOR) and World Agroforestry (ICRAF) effectively merged
[c] On December 2018, Bioversity International and the International Center for Tropical Agriculture (CIAT) signed a memorandum of understanding to create an alliance

3.3.2 Public-Private Partnerships

The public sector has played an enormous role in agricultural innovations, particularly in the United States within the system of land-grant colleges and universities, whose mission, as set forth in the Morrill Acts of 1862 and 1890, was to focus on teaching practical agriculture, science, military science, and engineering. Land-grant universities developed strong breeding programs which became the source of germplasm for American farmers and local seed companies. Academic research was pivotal in the development of hybrid seed programs which, in the case of maize, resulted in the discovery of heterosis, the tendency of a crossbred individual (a "hybrid") to show qualities superior to those of either parent ("inbreds" in the case of maize). The advent of hybrids, with their homogeneous growth across large expanses of Midwestern lands, allowed for the mechanization of agriculture. The deployment of hybrids led to major changes in the American farming system: farm horses were replaced by tractors, and mechanical harvesting replaced labor-intensive hand picking of hybrid corn. The transition from varietal breeding to double-cross and eventually single-cross hybrids, combined with improvement in land management techniques, crop protection products, and fertilizer programs, resulted in a threefold increase in maize yields between the 1920s and 1980s (Crow 1998).

Today, public-private partnerships remain an essential foundation of innovation and scientific advances, but the balance of resources devoted to research and development is rapidly changing. While it is undeniable that the private sector benefits greatly from investments made by the public sector in fundamental research and the education of scientists, companies have committed enormous amounts of resources to early stage discoveries to better serve farmers and seek an edge over their competitors. Against a backdrop of proportionally more limited resources for the public sector and innovations which require sophisticated (and expensive) infrastructure, the private sector has invested massively in technology developments, with global private spending on agricultural R&D rising from $5.1billion in 1990 to $15.6 billion by 2014 (Fuglie 2016), whereas the public sector increased at a slower rate. In 1980, the ratio of private vs. public investments was pegged at 0.54; by 2011, it had increased to 0.81, coming close to parity (Pardey et al. 2015). But it is not only the absolute amounts invested which have evolved over time, it is also the types of investments, which are undergoing a dramatic transition, particularly in large-scale agriculture.

Large companies in the agricultural sector compete with each other to deliver ever better products and services to their customers, who are increasingly discerning and sophisticated in choosing which products allow for sustainable returns. With farmer productivity being the key driver of an industry, which has innovated at a remarkable clip for close to a century, investments in research and development by the private sector have become a condition of existence. With the collaboration of James Watson and Francis Crick, which in 1953 led to the description of the structure of the double helix of deoxyribonucleic acid ("DNA") (Watson and Crick 1953), plant agricultural research advanced to be conducted at the molecular level. The second half of the twentieth century produced new and powerful scientific technologies, specifically based on recombinant DNA techniques, genetic engineering, rapid gene sequencing, and synthetic biology.

At first, large-scale collaborations focused on elucidating the genetic material of several plant species. The Multinational Coordinated Arabidopsis Genome Project was first, including stock centers in the United States and in Europe orchestrated in the United States by the National Science Foundation. After 10 years, the global collaboration involving scientists from the United States, Europe, and Japan resulted in a near complete *Arabidopsis* (commonly called thale cress) genome, published in the year 2000 (The Arabidopsis Genome Initiative 2000). Collaborations between academics and scientists from government and private industry were frequent and productive. Databases were built and shared, including by the private industry, like the single nucleotide polymorphisms and small insertions and deletions database constructed from the Columbia and Landsberg *erecta* ecotypes and made available by Cereon Genomics as a service to the community (Ausubel 2000). This spawned a series of large-scale projects aimed at discovering the functions of the 25,000+ genes identified in *Arabidopsis thaliana* (Bevan and Walsh 2005).

In September of 2002, the National Science Foundation (NSF) announced the launch of the Maize Genome Sequencing Project, but it was not until 2009 that a multi-institutional effort, involving scientists at Washington University in St. Louis, the Cold Spring Harbor Laboratory in New York, the Arizona Genomics Institute, and Iowa State University, resulted in the publication of a series of papers in the journal *Science* revealing the DNA sequence of maize B73 (Schnable et al. 2009), while in Mexico, scientists published results derived from the ancient popcorn variety palomero (Vielle-Calzada et al. 2009).

But while these initial efforts were largely supported by the public sector, the many billions of genotypic and phenotypic datapoints collected today by seed companies, particularly in the development of hybrid crops, have forced them to invest in infrastructure that dwarfs that of most academic institutions. Together with the curation of proprietary germplasm collections by long operating seed companies, this has allowed the private sector to become the major driver of the record yields that are obtained today with hybrid seeds in the developed world. Even so, collaborations between the public and the private sector remain essential for progress to be made and new inventions to be developed in hybrid breeding and more so with varietal crops, where the germplasm held and developed by universities remains the prevalent source of seeds for many farmers. Furthermore, such collaborations enable the spillover of technological progress to crops relevant in developing countries for which seed companies have not typically developed specific cultivars.

3.3.3 The Age of Genetically Modified Organisms (GMOs), Technology Alliances, and Biotechnology Startups

With the advent of a rapidly cycling simple model plant such as *Arabidopsis thaliana*, where forward genetics allowed academics all over the world to characterize genes with putative functions in plant development and the protection of yield, interactions between the public and the private sector became frequent and sustained, directed toward the elusive goal of extracting more yield from crop plants under normal and stressed conditions through the application of recombinant techniques. Startup companies spun off from universities and corporations and attracted venture capitalists eager to engage in an industry which promised large revenue potential. Calgene, founded in 1980 in Davis, California, was one of the first to believe that genetic engineering could be applied successfully to plant agriculture. Calgene's first product, the Flavr Savr tomato, was engineered with an antisense gene to downregulate the enzyme polygalacturonase which participates in the softening of the fruit and makes it susceptible to being damaged by fungal infections and postharvest handling. As a result, Calgene's Flavr Savr tomato could be harvested later and withstood storage and transport much better than conventional tomatoes. Calgene was also engaged in the development of herbicide-tolerant cotton, and canola that was genetically engineered to produce laureate, a key raw material used in the manufacture of soap, detergent, and personal care products.

Plant Genetic Systems ("PGS"), founded in 1982 by Marc Van Montagu and Jeff Schell who, at the University of Ghent, Belgium, and the Max Planck Institute in Germany, respectively, had developed a vector system for transferring foreign genes into the plant genome by using the Ti plasmid of *Agrobacterium tumefaciens*, is credited for significantly advancing the field of plant biotechnology. PGS was the first startup company to engineer tobacco plants with a gene encoding the insecticidal protein from *Bacillus thuringiensis* (Bt). DNA Plant Technology ("DNAP") was founded in 1981 by William R. Sharp and David A. Evans, in New Jersey, before moving to California "to develop tastier, value-added plant-based products for industrial and consumer markets"[3] through the use of advanced plant breeding systems, tissue culture, and later transgenic techniques. And with the advent of *Arabidopsis* genetics, companies were founded to explore gene structure-function relationships. Companies like Ceres Inc., founded in 1996 to explore gene expression patterns and functions; Mendel Biotechnology, Inc., founded in 1997 to understand gene expression through the study of transcription factors; and Paradigm Genetics, Inc., founded in 1997 to focus on functional genomics, became the new breed of innovators trying to revolutionize the industry. The new startup companies quickly formed strategic partnerships with major seed and trait developers such as Monsanto, Bayer, and DuPont. The large investments made by these seed companies to finance collaboration with the startups, combined with the support of venture capitalists, allowed for a golden age of small plant biotechnology companies which emerged in Belgium, the San Francisco Bay Area, Southern California, the Research Triangle Park of North Carolina, and progressively elsewhere in locations associated with major university centers.

For the next 10 years, further innovations involving recombinant techniques led to ever more innovative companies. DNA shuffling (or directed molecular evolution), a method for in vitro recombination of homologous genes, was invented by Dr. Willem ("Pim") Stemmer while working at Affymax Inc. Stemmer's invention led to the foundation of Maxygen, Inc., a startup company, which further developed the technology and innovated in the field of agronomic trait discovery and optimization through its Verdia, Inc. subsidiary. Maxygen and Verdia[4] formed several alliances, including with Pioneer Hi-Bred, Syngenta, and Delta & Pine Land Company, to "provide proprietary product solutions to important commercial problems in plant-based businesses through the application of advanced trait optimization methods."

---

[2] Author MM was employed by Maxygen and Verdia, Inc.

3.4 Closed Innovation

During most of the twentieth century, inwardly focused research and development (R&D) in industry enabled substantial achievements and many commercial successes.

In many countries, academic scientists were at the time more interested in the process of scientific discovery to gain insights about the physical world than in putting such knowledge into practical use. At that time, it was common among academic camps to dismiss researchers inclined toward applied applications of science, on the ground that they were sold to corporate interests and therefore considered greedy.

Therefore, and with the benefit of hindsight, during most of the twentieth century, pursuing R&D within a firm was one of the few options available to those interested in practical applications of science, and it became the *de facto* dominant design among large, multinational companies. Internal R&D divisions were considered a strategic asset that needed to be carefully nurtured and shielded from the external world and competitors. The world knew little about corporate innovations until scientific developments became launched and commercialized in the form of products (Chesbrough 2003).

After the Second World War, the US government made increasingly large amounts of funding available, not only to government labs but also to the large number of independent universities being created. The availability of more research grants and scholarships enabled a dramatic increase of the pool of talent educated at the graduate level in many fields of science. In addition, during the 1990s, private funding for R&D activities in developed countries caught up to public R&D budgets. The combination of additional funding, both public and private, and a larger pool of talent led to the golden era of corporate labs in firms such as DuPont, Bell Laboratories, General Electric, IBM, RCA, Xerox in the United States, BASF, Bayer, Roche, Nestlé, Unilever, Siemens, and Shell, among many others in Europe, and "keiretsu" conglomerates such as Mitsui, Mitsubishi, Sumitomo, and Sanwa, among others, in Japan.

The mindset was one of "closed innovation", and was based on the principle that companies were, by and large, on their own in terms of developing the technologies needed to sustain and increase their market footholds. It is during this period that the unfortunate "not invented here" motto was coined, reflecting an inward stance where firms could simply not afford to rely on scientific developments created outside the corporation, be these of public or private nature. From a strategic standpoint, internal R&D capabilities were seen as an entry barrier to discourage potential competitors (Chesbrough 2003).

Closed innovation created a subtle, and sometimes not so subtle, tension between research and development activities. For instance, while scientists in centralized, corporate research labs were driven to move into the next wave of innovative projects, instead of delving deeper into commercially relevant work, researchers in development felt the pressure to find out more about how to translate such research outputs into commercial products. In addition, while research usually represents an SGA (selling, general, and administrative) cost item from a budgetary perspective, development is typically allocated to business units and structured as part of profit and loss statements. Therefore, a functional and financial disconnect between research and development activities became more prevalent in many large corporate R&D organizations (Chesbrough 2003).

In more recent years, the closed innovation mindset has found itself increasingly at odds with the way that knowledge currently flows. There are several aspects which have severely undermined and challenged closed innovation, the most relevant being:

1. The availability of an increasingly more mobile pool of skilled researchers. Companies can build upon others' investments in a more effective way than before, for instance, through hiring skilled staff from other companies or through contracting consultants who provide advanced expertise in specific areas.
2. The availability of venture capital, which enables startups to pursue both research and development activities in a nimbler way than larger corporations. Venture capital has enabled independent startups within the agricultural domain with the funding needed to make significant and fast progress. For instance, Inari, a next-generation seed company which develops traits and cutting-edge technologies based on gene editing, raised $89 million in its series C financing round,[5] whereas Provivi, a company developing novel methods to control agricultural pests using naturally occurring pheromones, has raised over $36 million through its series B financing round.[6]
3. The increasing availability of external suppliers of R&D, both in numbers and in terms of the expertise they provide. For instance, the success in the ability to develop commercially effective Bt genes (used to develop biotech seeds with intrinsic resistance to insect damage) by Dow (now Corteva Agriscience) and Monsanto (now Bayer) can be traced back to the access of extensive libraries of Bt genes developed by smaller, specialized companies with a strong expertise in the field, such as Mycogen[7] and Ecogen,[8] respectively. While the development of transgenic drought traits has fallen short of expectations, with only Monsanto's DroughtGard™ hybrid maize reaching project launch status, 20 years of R&D and collaborations with startup companies such as Mendel Biotechnology, Ceres Inc., or Evogene have significantly contributed to increasing the scientific community's general knowledge of the molecular basis of drought resistance in plants (Nuccio et al. 2018). In recent years, the trend of favoring early stage discovery at startup companies has become the mainstream, since corporate venturing has allowed a large number of startup companies to be created not only in

the United States but also globally, some of which have become competitive in research domains such as gene editing and digital agriculture, among others. In 2017 alone, agriculture-related startups raised over $10 billion in venture-backed funding, and one of them, Indigo, has already reached unicorn[9] status.

4. The limited predictability in the usefulness of the extensive intellectual property portfolios companies have built over the years for the development of successful new products. A case in point is that less than 10% of Procter & Gamble's patents were used by any of its businesses for product development (Sakkab 2002). While patents can represent an effective trade currency to obtain freedom to operate (e.g., through cross-licensing) and keep litigation at bay, and if one excludes the small number of patents which sustain product launch and commercialization, the ability for a patent owner to translate a large number of them into actual value in the market is, at best, questionable. It is thus likely that making such intellectual property available to other organizations could increase the odds of deriving tangible value from it.

3.5 Open Innovation

Agriculture and agrifood companies of all sizes are currently beset with increasing competition and technological complexity, rising R&D costs, shortened product life cycles, increased expectations from customers, more complex and expensive regulatory processes, and industry consolidation. With improved institutional markets and stronger intellectual property rights, venture capital and technology standards have enabled organizations to trade their knowledge and ideas.

The challenge facing the domain of agriculture and agrifood systems is daunting, with current projections predicting the Earth's population ballooning to 9.7 billion by 2050, which is only 30 crop growing cycles away from today in the mid-2020. Though certain vegetable crops sustain more than one growing cycle per year, and while the climate in some countries such as Brazil and Paraguay also allows a second growing cycle of main crops each year (safrinha), by and large, the global commodities (maize, rice, soybeans, wheat) are only harvested once a year.

To place this challenging problem in context, humankind is going to have to produce as much food over the next 30 years as during the previous 10,000 years, with the added complexity of a reduction in available land per capita; the unpredictability brought about by climate change; an increasing concern by the public about the way crops and food are being produced, packaged, distributed, and commercialized; and the unsustainable fact that a large proportion of the food produced is being wasted.

[3] A privately held startup company with a current valuation of at least US$1 billion is referred to as a unicorn.

One consequence of the above-described conundrum is that even for large, multinational companies, the problem to solve is so large that it becomes untenable to maintain in-house all the expertise, infrastructure, and human talent needed to cover the continuum of activities ranging from fundamental research to those development activities leading to commercial launch inside the organization. Moreover, as described in Frans Johansson's fascinating book *The Medici Effect*, true innovation has the stubborn habit of coming from beyond your realm of expertise and from what he refers to as the "intersection of ideas."

In today's environment, regardless of their size, firms are quickly evolving from a closed into an open innovation mindset to remain competitive, since the previously successful closed mindset has become a factor stifling progress and reducing the likelihood of innovation success.

Henry Chesbrough from the University of California at Berkeley, who first coined the term open innovation (referred to thereafter as "OI" through this chapter), defines it as "a distributed innovation process based on purposively managed knowledge flows across organizational boundaries, using pecuniary and non-pecuniary mechanisms in line with the organization's business model" (Chesbrough and Bogers 2014). The main tenet of OI is that, in order to thrive in an increasingly competitive market and in a more complex society, firms should open up their boundaries and leverage inflows and outflows of knowledge and technology. In other words, companies should not only establish outside-in channels to source external knowledge but also develop inside-out processes to leverage external paths to novel markets which would otherwise remain off-limits. Figure 3.1 provides a current working model of open innovation. One of the major departures from a closed model is that companies practicing OI explicitly feed their innovations funnels with both, in-house and external developments. This allows them to evolve from the previous "not invented here" syndrome to a "proudly invented elsewhere" framework. In addition, OI enables not only the exploitation of current, known markets but also the exploration of novel ones, which brings about accelerated benefits for society at large, since under a closed model such developments and value creation activities would either never occur, or would take place several years later. Another benefit of OI, when it is linked with corporate entrepreneurship, is that OI efforts can create encouragement, opportunities, and incentives for staff development.

One of the underpinnings of OI is the realization that the knowledge needed to innovate is becoming more widely distributed in the economy and that there is an increased scattering of the sources and providers of such knowledge. Under these premises, the two main working models of OI, namely, outside-in and inside-out innovation, where firms are open to adopting all kinds of external inputs and contributions, while simultaneously allowing some of their technological assets to be used by others, makes sense. In coupled modes of open innovation, outside-in and inside-out modes are combined.

OI can be implemented through several approaches such as in- and out-licensing, spin-ins (or acquisitions) and spin-outs, co-conception and co-development,

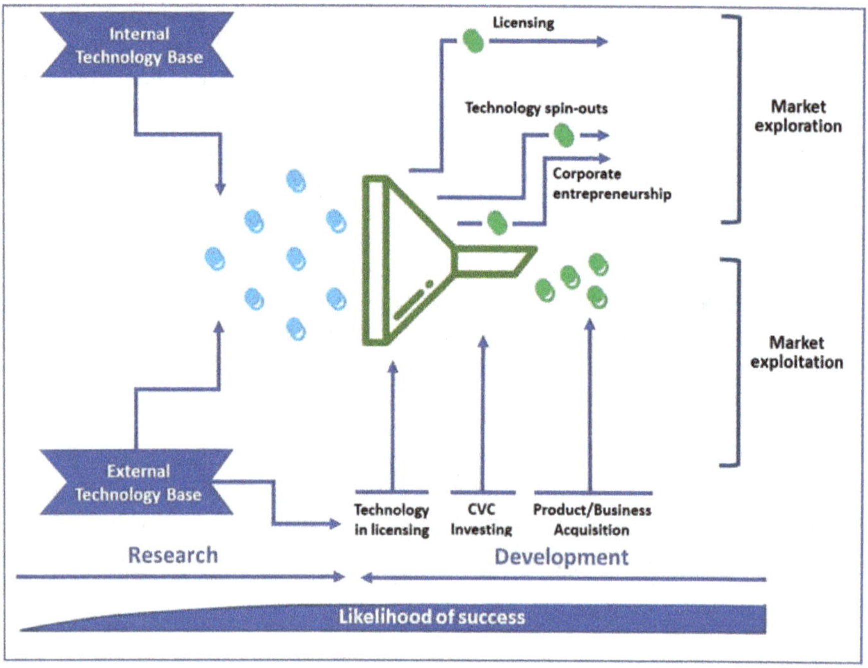

Fig. 3.1 Open innovation

research joint ventures, crowdsourcing, community innovation, and online market places. In some cases, a company implementing OI can also enable employees to exploit a given technology in noncompeting areas or in areas not yet ripe for product development. Furthermore, OI efforts can develop hitherto non-exploited markets through corporate entrepreneurship efforts, as already mentioned. Regardless of the approach, all forms of OI enable a quicker and nimbler flow of knowledge between different actors at the supply and demand ends of the market.

From a conceptual innovation standpoint, OI embraces the "jobs to be done" perspective described in Chap. 1 in this volume, since it aims to reduce satisfaction gaps which create both frustrations in customer and business opportunities for firms that would likely otherwise go unnoticed, or be exploited by competitors.

OI represents the explicit recognition that even for very large companies, it is not in their best interest to attempt to internally address all R&D aspects to remain competitive but also that for some of their R&D outcomes, there are unchartered markets previously untapped from which value can be captured by making such outcomes available to other organizations.

In order to understand how diverse organizations create the paths to both engage and benefit from OI, authors such as Linus Dahlander and David Michael Gann have elaborated a capability perspective, which considers the organizational culture and the mechanisms that need to be developed to create innovations (Dahlander and Gann 2010). Differences in capabilities will determine the extent to which a firm implements OI and its ability to create value. Furthermore, a capabilities' perspective helps to design OI systems able to become a competitive advantage in a world where technical knowledge doubles every 2 years and where more and more knowledge is either easily distributed or is made available as a public good in the public domain. At the most fundamental level, OI is embedded in the notion that the sources of knowledge for innovation are widely distributed in the economy.

Embracing OI modifies the mindset that companies take to the development of new products. Christensen (2006) points out that to maximize the benefits of OI, companies need to not only possess deep expertise and technological ability but also combine the ability to integrate sourced innovations or intellectual property, manage relationships with different partners, and sometimes adopt a dual, possibly simultaneous role, as both provider and seeker of technology.

3.5.1 *Open Innovation and Intellectual Property*

Open innovation is not open source. Patents are important as they encourage firms to invest in R&D and to disseminate technical knowledge within the economy. They are granted to a holder by the appropriate national or regional authority and provide limited rights to the patent holder to exclude others from commercially exploiting a given invention over a set period of time, unless the patent holder authorizes them to do so.

Patents are present throughout all the steps of the open innovation process as a structuring element. Initially, they facilitate knowledge inflows through signaling and in-licensing contracts. Then they help to coordinate the production of knowledge by collaborators and competitors, starting with their use as bargaining chips and ending with their potential as contract enforcers. Finally, they help manage the outflows of knowledge and the distribution of profits by enabling either out-licensing strategies or by ensuring the maintenance of exclusivity and its associated profits. As such, in open innovation, firms are not only producers but also active users of patents and other means of intellectual property rights (IPRs) (Pénin and Neicu 2018).

A recent study analyzed the 100 largest R&D spenders among Swedish national and binational firms, as well as the 100 largest R&D spenders among Swedish small and medium enterprises, to shed light on the relationship existing between patenting and OI. Patenting with protection (technology protection and safeguarding, blocking competitors from certain technology areas, obtaining freedom to operate (FTO), and creating retaliatory power) and bargaining motives (enabling licensing, cross-licensing, and R&D collaborations) were positively related with the importance of open innovation strategies. However, the positive relationship between protection motives and open innovation was stronger than that between bargaining motives and open innovation. According to the study, patenting acts as an enabler of open innovation for the individual innovating firm (Holgersson and Granstrand 2017).

Though the abovementioned study used patents, there are additional mechanisms of intellectual property protection. A large study conducted in Spain with over 2,800

SME firms concluded that, while SMEs do not directly benefit from open innovation or from patenting in the same way as larger firms, they nevertheless profit from alternative ways of protecting the intellectual property associated with their proprietary technologies (Brem et al. 2017).

Regardless, the proper management and availability of IPRs in a general sense exert a positive influence on OI efforts.

3.6 Open Innovation Comes of Age in Crop Genetics

The expenses and complexities associated with the development of transgenic traits have kept the development of GMO crops out of reach of most but for the largest companies supplying large scale agriculture with maize and soybean seeds. Because model plant systems were, by and large, unable to predict the effects of multigenic traits affecting yield and other quantitatively inherited traits of importance to farmers, the seed industry relied on an innovation model that became increasingly closed in areas such as transgenic trait development and predictive breeding. However, with the recent consolidation of the industry and the resultant restructuring and focus on product development, a compelling argument can be made that the next decade will see a more open form of innovation, centered around early stage discovery. The phenomenon is not unlike what the pharma industry experienced over the past two decades, with the integration of technologies developed by startups becoming their major source of innovation.

Large, medium, and small seed and plant breeding companies have much to gain from embracing an open innovation model, due to the types of innovations which will be needed by the farming community in the next decade or two and the cost of developing these innovations. Firms in the precision farming sector, which rely on the confluence of many disciplines from the high tech, data management, computer vision, and agronomy sectors, will benefit as well.

The successful deployment of new technologies is dependent not only on farmer adoption but also on the appropriate regulatory framework, which in turn is contingent on the acceptance of such technologies by consumers, the general public, and society at large. Past experience with transgenic traits has shown that in the absence of obtaining a social license to practice the new technologies, consumer acceptance may be difficult or even impossible to obtain. This argues for the large developers of genetics in commodity crops to make their technologies and know-how available broadly to stimulate entities more focused on academic pursuits and the development of public goods to become early adopters and independent champions of such technologies. For example, smaller fruit and vegetable seed companies and nonprofit organizations can harness CRISPR-Cas9 genome editing and associated technologies to develop disease resistance traits creating the potential for more sustainable, nutritious, and affordable products, for both developed and developing countries. To enable such a model and also advance its philanthropic aims, DuPont Pioneer (now part of Corteva Agriscience) developed, in 2017, a joint licensing framework with the Broad Institute of Harvard University and the Massachusetts Institute of Technology (MIT) to provide access to CRISPR-based gene editing technologies in plant agriculture. Under the joint licensing structure, the parties made their combined portfolio of over 40 patents available for free to support research conducted by academics, governmental agencies, and nonprofit organizations, as well as for the development of agricultural products destined to smallholder farmers, while also retaining the freedom to license their individual patent portfolio to others. For other licensees, they apply financial terms that are adapted to the licensees' size and revenue prospects.

The long and arduous path that often accompanies the establishment of research and commercial licensing agreements provides additional support for a more open form of innovation. Early stage technologies are highly uncertain, and many remain untested, either due to failure to reach a satisfying win-win situation during negotiations or simply because the negotiation process constitutes a hurdle that the parties are not willing to face. As Fig. 3.1 shows, the likelihood of success of technologies at a very early stage of development is, at best, slight. Settling for simpler, research-only licensing relationships can open the floodgates that keep less mature technologies from being tested while deferring the negotiation of commercial terms to a time when initial proof of concept is confirmed. Under a closed innovation mindset, such technologies might never be tested. Though this approach might not be adequate for all technologies, it represents an option that both sides of a negotiation might consider.

A further reason for encouraging a more open form of collaboration and innovation is the highly interconnected and interdependent nature of technology developments among many participants in the agricultural technology development sector. Technological and commercial developments rarely happen in isolation but rather derive from the development of networks of agricultural technologies, in which components interact and co-evolve to become increasingly interdependent. One of the reasons for such interdependency is the establishment of separate dominant designs, which create network externalities, complementing each other and lowering the costs of absorbing and mainstreaming new developments.

This creates multitiered hierarchical structures where end-user technologies become dependent upon infrastructure technologies which underlie the solutions stack. This in turn leads to the establishment of platform technologies of increasing relevance in agriculture, not only in commercial terms but also because of their ability to create additional value for customers and to integrate digital, biological, and financial solutions in a more efficient and/or affordable way.

While startups and academics have vast innovative abilities and venture capitalists have resources that can be assigned to the development of new technologies, only the large, international seed companies have the scale and the development resources to bring some of these innovations to market in production agriculture.

3.7 The Role of Venture Capital (VC)

The 1980s and 1990s saw a flurry of entrepreneurs developing their proprietary plant biotechnology innovations through newly formed startup companies. Venture capitalist investment followed the successful strategic partnerships that startups formed with major seed companies, and the first wave of successful exits resulted in acquisitions by the multinationals who became keen to internalize the innovative discovery programs of the startups. However, by 2010, plant biotechnology startups had thinned, venture capitalists had become wary of the long development timelines associated with transgenic crops, and the multinationals had taken advantage of the historically high seed prices of 2008, 2011, and 2012 to invest in and develop their own formidable internal programs. Another realization started to set in: beyond the initial biotech traits conferring herbicide tolerance and insect resistance, agricultural biotechnology's holy grail of increased drought tolerance, yield, and fertilizer use efficiency would require harnessing complex multigenic traits. Model plant systems were largely inadequate to tease out the complex requirements that would result in a successful trait without pleiotropic effects. Interests shifted, and new areas of innovation started to make it to the foreground. The soil and plant microbiomes became an innovation target.

For millions of years, plant and soil microbes have coexisted and benefitted from each other. Microbes decompose organic matters and make nutrients such as nitrogen and phosphorus available to the plants, and the plants supply the carbon needed by the bacteria to survive. In 2009, Bayer Crop Science announced its new VOTiVO™ biological seed treatment, which makes use of beneficial *Bacillus firmus* bacteria to protect maize, soybeans, and cotton against nematodes. In 2012, Syngenta acquired the startup Pasteuria Bioscience, with which it had partnered since 2011. This was followed in 2013 by the launch of Clariva pn, Syngenta's first biological nematicide seed treatment, which contains spores of *Pasteuria nishizawae*, an obligate parasitoid of soybean cyst nematodes. The race was on! Bayer acquired California-based AgraQuest in 2012; Monsanto announced a deal with Torrey Pines, CA-based Synthetic Genomics and formed the BioAg Alliance to commercialize microbial products for agriculture with Denmark based Novozymes, the world's largest enzyme producer, both in 2013. The next few years were a golden age for VC and startups in the biological space. Bayer purchased Argentinian Bioagro Group in 2014 and announced a research collaboration with St. Louis, MO-based Elemental Enzymes in 2015. DuPont acquired Taxon Bioscience, a Tiburon, CA-based startup focusing on industrial microbes; and Dow AgroSciences entered a collaboration with UK-based Synthace Ltd., both in 2015. In 2016, Monsanto's corporate venture group invested in California-based Pivot Bio and North Carolina-based AgBiome; in 2017 it led a financing round in St. Louis-based NewLeaf Symbiotics. In 2017, Bayer announced a new partnership with Ginkgo Bioworks to develop microbial products that would stimulate nitrogen fixation in plants, and DowDuPont announced a collaboration with Arysta LifeScience and a multi-year collaboration with Israel-based Evogene, to develop microbial seed treatments for maize.[10]

Another area of intense innovation and startup activity in the last 5–10 years is that of precision agriculture. Further details on this topic are provided by Sonka in Chap. 8 of this volume. At the convergence of big data management, satellite imagery, remote sensing, computer vision, and precision input application technologies, startups and venture capitalists are hustling to develop the technologies that will make farming more data driven, productive, and sustainable. Precision agriculture is not a new concept, as farmers adopted GPS-connected equipment in the 1990s to increase productivity; but with the advent of remote sensing, unmanned aerial vehicles (UAVs, aka drones), high-speed internet, and other technological advancements, precision agriculture is becoming the mainstream.[11] As it happened with other technologies, the large seed companies invested heavily to develop their digital agricultural services' platforms. The most notable exits in this space are the Monsanto's acquisition of the Climate Corporation in 2013 and DowDuPont's acquisition of Granular in 2017.

Today's farmers have access to an enormous amount of data. But data without integration and management recommendations are of little use. Some of the challenges facing precision agriculture are the integration of big data into a single platform, interpretation of remote sensing and scouting data through crop consultants or apps that make use of artificial intelligence, and the implementation of recommendations about planting decisions and application of inputs across entire farms.

But the area which has perhaps attracted the most interest from scientists, investors, and the public alike is that of genome editing. With the discovery of CRISPR-Cas9 genome editing technology in 2012, published almost simultaneously by scientists at the Vilnius University, Lithuania (Gasiunas et al. 2012), on one hand, and by scientists at the University of California at Berkeley, the University of Vienna, and Umea University (Jinek et al. 2012) on the other, the field immediately triggered a flurry of entrepreneurial activity, with the founding of multiple startups and massive investments by the venture capital community. The world of plant genetic innovations was once again coming into focus. While genome editing wasn't new in itself, since it had been pioneered through the use of zinc finger nucleases as early as the 1980s and by using meganucleases and TALENs thereafter, CRISPR technology made the precise editing of genomes an affordable and facile endeavor. Before CRISPR, most companies had to resort to using transgenes to bring genetic improvements to market, a process which took in average 13 years and cost $130MM for existing products[12] and was increasingly becoming more costly and taking longer, as the regulatory path became gradually more complicated. This limited transgenic improvements only to major crops such as maize and soybeans, where companies could envision making a return on their investment. With CRISPR technology and the prospects of an appropriate regulatory framework in the United States and other countries recognizing that CRISPR-edited crops are not transgenic crops, the field opened widely to academic researchers and startups.

There is no doubt that AgTech innovations are experiencing a golden age and that the current venture system represents its lifeblood. Investments funding AgTech startups reached an all-time high in 2015, with $4.6 billion committed, followed by $4.2 billion in 2016.

3.8 Mergers and Acquisitions

Consolidation in the seed and crop protection industry over the past decades led to the formation of large organizations uniquely positioned to exploit increasingly complex molecular breeding and transgenic trait development programs. During the golden years of high seed prices, substantial capital investments into large internal discovery engines were made by the multinationals.

Nevertheless, while transgenic input traits, including herbicide tolerance and insect control traits, translated into increased revenues for the seed companies and efficiency gains for farmers, investments committed to the development of transgenic agronomic traits such as intrinsic yield increase, or increased fertilizer efficiency, have yet to deliver value.

With yields increasing steadily and seed stocks in the developed world attaining record highs, commodity prices collapsed and so did the growth of company revenues. This led to yet another round of consolidation resulting in the acquisition of Syngenta by ChemChina, the merger of equals between Dow and DuPont, and the acquisition of Monsanto by Bayer.

The merged firms have consolidated development pipelines which, for years to come, will require large amounts of resources to drive product concepts toward commercial launch. Combined with the need for these firms to reduce their marginal costs, this could put pressure on their investments in early stage discovery, which will result in an opportunity for the startup sector and academics alike to regain an important role in supplying the industry with the cutting-edge technologies of tomorrow.

3.9 Closing Comments

Collaborations between the private and public sector have brought about steadily increasing yields. Incredible productivity gains have been obtained by a sophisticated industry that has successfully incorporated advanced technologies. Since the advent of hybrid maize, yields in the United States have increased sevenfold! Corteva Agriscience has built a maize germplasm collection that goes back over 90 years, with phenotypic data that was collected from the start and associated genotypic data built over time. Such continuity in data collection and the coherence of the germplasm pools are unmatched in the private or the public sectors. Combined with an ability to precisely correlate phenotype to genotype for the past two decades, the accumulated data allows for increasingly accurate predictions in support of the breeding process, allowing modern breeders to cross inbreds chosen from the germplasm pools to obtain high-performing hybrids that respond to specific biotic and abiotic stress conditions while maximizing yield for the targeted maturity zone. That is the strength of the large agricultural input multinationals. But we are entering an era of complementarity with developers of early stage technologies, whether they are in the public sector or at startups, where the diverse models of open innovation are expected to play a larger role than in the past. The convergence of innovators in a highly technological sector of enormous breadth and scale has the potential to shape the future of agriculture into one that can feed the world's nine billion people in a sustainable and environmentally friendly way.

Acknowledgments Innovation is about people's connections, and we would like to wholeheartedly thank Dr. David Meyer (Corteva Agriscience) for connecting us. Without David's willingness and generosity to put us in touch, this chapter would not have been possible. Hugo Campos acknowledges the financial support of the CGIAR Research Program on Roots, Tubers and Bananas (RTB), supported by CGIAR Trust Fund contributors (https://www.cgiar.org/funders/), and is also very grateful to the support of a Bill & Melinda Gates Foundation investment (OPP1213329) awarded to the International Potato Center.

We would also like to thank Prof. Henry Chesbrough (University of California at Berkeley) for granting permission to use, and modify, Fig. 3.1.

References

Ausubel FM (2000) Arabidopsis genome. A milestone in plant biology. Plant Physiol 124:1451–1454

Bernardo R, Yu J (2007) Prospects for genome-wide selection for quantitative traits in maize. Crop Sci 47:1082–1090

Bevan M, Walsh S (2005) The Arabidopsis genome: a foundation for plant research. Genome Res 15:1632–1642

Bocquet-Appel JP (2011) When the world's population took off: the springboard of the Neolithic demographic transition. Science 333(6042):560–561

Brem A, Nylund P, Hitchen E (2017) Open innovation and intellectual property rights. Manag Decis 55(6):1285–1306

Chen K, Wang Y, Zhang R, Zhang H, Gao C (2019) CRISPR/Cas genome editing and precision plant breeding in agriculture. Annu Rev Plant Biol 70:667–697

Chesbrough HW (2003) Open innovation. Harvard Business School Press, Boston, 227p

Chesbrough HW, Bogers M (2014) Explicating open innovation: clarifying an emerging paradigm for understanding innovation. In: Chesbrough HW, Vanhaverbeke W, West J (eds) New frontiers in open innovation. Oxford University Press, Oxford, United England, pp 1–37

Christensen JF (2006) Wither core competency for the large corporation in an open innovation world? In: Chesbrough HW, Vanhaverbeke W, West J (eds) Open innovation: researching a new paradigm. Oxford University Press, Oxford, UK

Crow JF (1998) 90 years ago: the beginning of hybrid maize. Genetics 148(3):923–928

Dahlander L, Gann DM (2010) How open is innovation? Res Policy 39(6):699–709

Diamond J (1997) Guns, germs, and steel: the fates of human societies. W.W. Norton & Company. 480 p. ISBN 978-0-393-03891-0. OCLC 35792200

Fuglie K (2016) The growing role of the private sector in agricultural research and development world-wide. Glob Food Sec 10:29–38

Gasiunas G, Barrangou R, Horvath P, Siksnys V (2012) Cas9 crRNA ribonucleoprotein complex mediates specific DNA cleavage for adaptive immunity in bacteria. Proc Natl Acad Sci U S A 109:2579–2586

Gilmour AR, Cullis BR, Verbyla AP (1997) Accounting for and extraneous variation in the analysis of field experiments. J Agric Biol Environ Stat 2(3):269–293

Helentjaris T, Slocum M, Wright S, Schaefer A, Nienhuis J (1986) Construction of genetic linkage maps in maize and tomato using restriction fragment length polymorphisms. Theor Appl Genet 72(6):761–769

Holgersson M, Granstrand O (2017) Patenting motives, technology strategies, and open innovation. Manag Decis 55(6):1265–1284

Horsch, R. B., Fry, J. E., Hoffmann, N.L., Eicholtz, D., Rogers, S. G,.Fraley, T. A. 1985. A simple and general method for transferring genes into plants. Science 227: 1229–1231

Jinek M, Chylinski K, Fonfara I, Hauer M, Doudna JA, Charpentier E (2012) A Programmable Dual-RNA–Guided DNA Endonuclease in Adaptive Bacterial Immunity. Science 337(6096):816-821

Kasha KJ, Kao KN (1970) High frequency haploid production in barley (*Hordeum vulgare* L.). Nature 225:874–876

Kavanagh PH, Vilela B, Haynie HJ, Tuff T, Lima-Ribeiro M, Gray R, Botero C, Gavin M (2018) Hindcasting global population densities reveals forces enabling the origin of agriculture. Nat Hum Behav 2:474–484

Maisels CK (1993) The near east: archaeology in the cradle of civilization. Routledge Press, 241 p. isbn:978-0-415-04742-5

Mendel G (1866) Versuche über Plflanzenhybriden. Verhandlungen des naturforschenden Vereines in Brünn, Bd. IV für das Jahr, Abhandlungen, 3–47

Nuccio ML, Paul M, Bate NJ, Cohn J, Cutler SR (2018) Where are the drought tolerant crops? An assessment of more than two decades of plant biotechnology effort in crop improvement. Plant Sci 273:110–119

Pardey P, Chan-Kang C, Dehmer S, Beddow J (2015) InSTePP's international innovation accounts: research and development spending, version 3.0. InSTePP, University of Minnesota, St. Paul

Paterson AH, Lander ES, Hewitt J, Peterson S, Lincoln SE, Tanksley SD (1998) Resolution of quantitative traits into Mendelian factors by using a complete linkage map of restriction fragment length polymorphisms. Nature 335:721–726

Pénin J, Neicu D (2018) Patents and open innovation: bad fences do not make good neighbors. J Innov Econ Manag 25:57–85

Platt DE et al (2017) Mapping post-glacial expansions: the peopling of Southwest Asia. Sci Rep 7:40338. https://doi.org/10.1038/srep40338. ISSN 2045-2322

Podlich D, Cooper M (1998) QU-GENE: a simulation platform for quantitative analysis of genetic models. Bioinformatics 14(7):632–653

Sakkab N (2002) Connect & develop complements research and develop at P&G. Res Technol Manag 45(2):38–45

Schnable PS, Ware D, Fulton RS, Stein JC, Wei F, Pasternak S, Liang C, Zhang J, Fulton L, Graves TA, Minx P, Reily AD, Courtney L, Kruchowski SS, Tomlinson C, Strong C, Delehaunty K, Fronick C, Courtney B, Rock SM, Belter E, Du F, Kim K, Abbott RM, Cotton M, Levy A, Marchetto P, Ochoa K, Jackson SM, Gillam B, Chen W, Yan L, Higginbotham J, Cardenas M, Waligorski J, Applebaum E, Phelps L, Falcone J, Kanchi K, Thane T, Scimone A, Thane N, Henke J, Wang T, Ruppert J, Shah N, Rotter K, Hodges J, Ingenthron E, Cordes M, Kohlberg S, Sgro J, Delgado B, Mead K, Chinwalla A, Leonard S, Crouse K, Collura K, Kudrna D, Currie J, He R, Angelova A, Rajasekar S, Mueller T, Lomeli R, Scara G, Ko A, Delaney K, Wissotski M, Lopez G, Campos D, Braidotti M, Ashley E, Golser W, Kim H, Lee S, Lin J, Dujmic Z, Kim W, Talag J, Zuccolo A, Fan C, Sebastian A, Kramer M, Spiegel L, Nascimento L, Zutavern T, Miller B, Ambroise C, Muller S, Spooner W, Narechania A, Ren L, Wei S, Kumari S, Faga B, Levy MJ, McMahan L, Van Buren P, Vaughn MW, Ying K, Yeh CT, Emrich SJ, Jia Y, Kalyanaraman A, Hsia AP, Barbazuk WB, Baucom RS, Brutnell TP, Carpita NC, Chaparro C, Chia JM, Deragon JM, Estill JC, Fu Y, Jeddeloh JA, Han Y, Lee H, Li P, Lisch DR, Liu S, Liu Z, Nagel DH, McCann MC, SanMiguel P, Myers AM, Nettleton D, Nguyen J, Penning BW, Ponnala L, Schneider KL, Schwartz DC, Sharma A, Soderlund C, Springer NM, Sun Q, Wang H, Waterman M, Westerman R, Wolfgruber TK, Yang L, Yu Y, Zhang L, Zhou S, Zhu Q,

Bennetzen JL, Dawe RK, Jiang J, Jiang N, Presting GG, Wessler SR, Aluru S, Martienssen RA, Clifton SW, McCombie WR, Wing RA, Wilson RK (2009) The B73 maize genome: complexity, diversity, and dynamics. Science 326(5956):1112–1155

Tallavaaraa M, Eronen JT, Luoto M (2018) Productivity, biodiversity, and pathogens influence the global hunter-gatherer population density. Proc Natl Acad Sci U S A 115:1232–1237

The Arabidopsis Genome Initiative (2000) Analysis of the genome sequence of the flowering plant Arabidopsis thaliana. Nature 408:796–815

Vavilov NI (1951) The origin, variation, immunity and breeding of cultivated plants (translated by K. Starr Chester). Chron Bot 13:1–366

Vielle-Calzada JP, de la Vega OM, Hernández-Guzmán G, Ibarra-Laclette E, Alvarez-Mejía C, Vega-Arreguín JC, Jiménez-Moraila B, Fernández-Cortés A, Corona-Armenta G, Herrera-Estrella L, Herrera-Estrella A (2009) The Palomero genome suggests metal effects on domestication. Science 326:1078

Watson JD, Crick F (1953) The structure of DNA. Cold Spring Harb Symp Quant Biol 18:123–131

Weiss E, Kislev M, Hartmann A (2006) Autonomous cultivation before domestication. Science 312:1608–1610

Wolter F, Schindele P, Puchta H (2019) Plant breeding at the speed of light: the power of CRISPR/Cas to generate directed genetic diversity at multiple sites. BMC Plant Biol 19:176

4

Adaption and Public Reluctance: Theories and Modes of Thinking

Cees Leeuwis and Noelle Aarts

4.1 Introduction

While many scholars and practitioners use the term 'agricultural development', the meaning of the concept is ambiguous, as there exist different views on the desired model of agriculture. For example, some associate 'agricultural development' with an 'agro-ecological' future, while others make a plea for 'sustainable intensification'. Regardless of such view, agricultural development is somehow associated with changes in agricultural practices, including the so-called adoption of technological and other types of innovation that are congruent with a proposed agricultural development trajectory. The term adoption has a long tradition and has been used in many studies (Rogers 1962; Loevinsohn et al. 2012, 2013). Frequently, such studies assess the adoption of new practices and technologies as part of an effort to determine the success or failure of development interventions. At the same time, several scholars have pointed to difficulties with regard to such a use of the term 'adoption' in research and development practice (Loevinsohn et al. 2012, 2013; Glover et al. 2016). It is argued that it is far from clear when someone can be regarded as having actually 'adopted' a practice or technology; should we speak of adoption when someone uses a new practice on a small portion of a farm, for a short period of time, or for merely opportunistic reasons (e.g. to get access to a service or network)? Another critique is that the term obscures the socio-technical character of change, as well as emergent forms of creolisation, making and redesign that frequently

C. Leeuwis (✉)
Knowledge, Technology and Innovation group at the Section Communication, Philosophy and Technology, Wageningen University, Wageningen, The Netherlands
e-mail: cees.leeuwis@wur.nl

N. Aarts
Socio-Ecological Interactions group at the Institute for Science in Society, Radboud University, Nijmegen, The Netherlands
e-mail: noelle.aarts@ru.nl

occur when farmers engage with new technologies (Glover et al. 2016). In other words, that what is supposedly adopted may actually have changed and become contextually adapted. Such critiques are valid and indeed render the concept of 'adoption' to be rather ambiguous.

In this chapter, we do not focus on the issue of how to precisely define or measure adoption, nor on the processes of adaptation that are likely to be involved. Instead, we focus on theories and modes of thinking that have been used to explain and understand why people do or do not 'adopt' a novel practice. This regardless of how sizable, durable or adaptive such changes may be. Such theoretical explanations include two interrelated dimensions: (a) factors or variables that are seen to influence whether people adopt or not and (b) the kinds of dynamic processes that are seen to be at work in shaping such factors. In relation to these, we will argue that adoption theories and technology uptake models have tended to regard adoption as a largely individual process and that they have thereby overlooked critically important interdependencies in the process of adoption. Subsequently, we complement individualist models for explaining adoption with sociological variables and then work towards an interactional perspective on the adoption process. Eventually, we will briefly highlight relevant conceptual and practical implications of our perspective with regard to some topical issues in development intervention, notably the idea of 'scaling' and the wish to support adoption and scaling with the help of information and communication technology for agriculture (ICT4Ag).

4.2 Adoption Seen as an Individual Argumentative Process

4.2.1 Everett Rogers' Seminal Work

In 1962, Everett Rogers made a synthesis of more than 500 studies in his famous book 'diffusion of innovations' in which adoption is a central theme. At this point in time, adoption is regarded as a process of individual decision-making, even though it is recognised that individuals are part of a broader system (Rogers 1962). Rogers argues that adoption (or rejection) of an innovation arises from a process in which individuals go through several stages (see Table 4.1). The process is essentially seen as a cognitive process in which various sources of information may play an important role. In line with this, provision of particular information was seen as a key activity in supporting individuals in the process of arriving at an adoption or rejection decision (Van den Ban 1963; Van den Ban and Hawkins 1996).

According to Rogers, the way an adoption process evolves is further shaped by the antecedents of the individual involved (his or her values, status, cosmopoliteness, etc.), his or her perception of the situation and the perceived characteristics of the innovation. This includes how individuals view the relative advantage of the innovation, the extent to which the innovation is regarded as compatible with existing practices and the perceived complexity of using the innovation in the context at

Table 4.1 The adoption process according to Rogers 1962 (and 1995) linked to information that is relevant per stage (Van den Ban 1963; Van den Ban and Hawkins 1996; Leeuwis 2004)

Stages in the adoption process (below in brackets: renaming of stage in Rogers 1995)	Nature of knowledge and information required and/or searched for
Stage 1: Awareness of the existence of a new innovation or policy measure (Knowledge)	Information clarifying the existence of tensions and problems addressed by the innovation or policy measure
Stage 2: Interest collecting further information about it (Persuasion)	Information about the availability of promising solutions that may be relevant in the prospective user context
Stage 3: Evaluation reflection on its advantages and disadvantages (Decision)	Information about relative advantages and disadvantages of alternative solutions
Stage 4: Trial testing innovations/behaviour changes on a small scale (Implementation)	Feedback information from one's own or other people's practical experiences
Stage 5: Adoption (or Rejection) applying innovations/behaviour changes (Confirmation)	Information reinforcing the adoption decision made

hand. Some of these perceptions are in fact reconfigured during the adoption process; based on information acquired, a person's perception of the situation and/or the perception of the relative advantages of the innovation may change.

Rogers seminal synthesis has been amended and developed further since its appearance in 1962 (see Rogers 1995). Moreover, the overall perspective and largely linear paradigm that it reflects have been criticised sharply (see, e.g. Leeuwis 2004).

4.2.2 Social-Psychological Follow-Up

Rogers' focus on individual decision-making has inspired many scholars to apply social-psychological theories and models to the issue of adoption (Taherdoost 2018). Such models include the theory of reasoned action (TRA, Fishbein and Ajzen 1975), the theory of planned behaviour (TPB, Ajzen 1985, 1991), different versions of the technology acceptance model (Davis et al. 1989; Venkatesh and Davis 2000), social cognitive theory (SCT, Rana and Dwivedi 2015) and uses and gratification theory (U&G, West and Turner 2010). Several such models were integrated in the unified theory of acceptance and use of technology (UTAUT) by Venkatesh et al. (2003).

While these models differ from each other in several respects, they tend to include a number of perceptual, cognitive and/or mental variables, often labelled as 'determinants' of adoption behaviour. Some key categories of determinants are listed below (see Table 4.2). As shown in Fig. 4.1, such determinants are often regarded as being influenced by personality traits and demographic variables.

4.3 Considerations Related to Interdependence: Towards an Interactional View on Reasons for (Non)adoption

The individualist perspective on adoption has been criticised for its simplistic notion of adoption (see the introduction), its pro-innovation bias, its linear connotations and its lack of attention for social and institutional dimensions of innovation (see Leeuwis 2004). Similarly, it has been argued that the kind of social-psychological theories referred to above overemphasise cognitive and rational processes and do not sufficiently acknowledge limitations in our cognitive capacities (Kahneman 2011) and the importance of heuristics, routine and/or impulsive triggers for behaviour and behaviour change (Petty and Cacioppo 1986; Baumeister et al. 2007). While these are all valid critiques, we would like to discuss less clearly elaborated shortcomings that arise from years of experiences in studying the uptake of technology. To highlight these, we continue to assume situations where individuals *are* indeed explicitly considering to some degree whether or not to adopt some kind of practice or technology. We first point to the issue of interdependence as an important omission in the thinking about adoption and then propose additional factors and variables that may explain why people do or do not adopt a new behaviour or technology.

4.3.1 The Issue of Interdependence

An issue that tends to be insufficiently acknowledged in the individualist behavioural perspective on adoption is that the performance of a specific behaviour is always linked with (and dependent on) the performance of other behaviours (Aarts 2018a). Several types of interdependence can be distinguished, labelled below as vertical, horizontal, intra-individual and time-related interdependencies.

Vertical interdependencies When considering whether or not to adopt a behaviour or technology, people frequently depend on the behaviours of dissimilar others. Farmers, for example, are embedded in several value chains and will consider a new practice in the context of what others in a value chain may do or not do. When considering whether or not to adopt a new crop variety, they may take into account whether or not local traders will actually be interested to collect and buy the produce, whether the agro-dealer will be willing to provide necessary pesticides on

Table 4.2 Overview of categories of determinants that are seen to influence adoption behaviour (similar terms used instead of 'adoption behaviour' include actual use, usage behaviour, behaviour, adoption)

Category of determinants (Similar terms used in literature)	Description
1. Intention (behavioural intention, intention to use)	In many models, it is assumed that adoption behaviour is preceded by an overall predisposition to perform the behaviour. This is the weighed resultant of several other components in this table (notably 2, 4 and 7)
2. Attitude (perceived usefulness, relative advantage, attitude towards use)	Intention is supposedly shaped by a variable that relates to an overall feeling that individuals have towards a behaviour or technology, which can be positive, negative or neutral. Attitude is seen to arise from two sub-variables, one relating to knowledge and the other to values (see 3. and 4.)
3. Knowledge (behavioural beliefs, outcome expectation, performance expectation, behavioural consequences, belief about outcomes)	Several models explicate that the attitudinal component (see 2.) is influenced by what actors perceive to be the consequences of adopting the behaviour: 'if I adopt this technology/behaviour, then X, Y, Z is likely to happen'. In essence, this is about predictive knowledge and understanding
4. Values (outcome evaluation, aspirations, goals)	Several models explicate that the likely consequences of adoption (see 3.) are weighed against people values, that is, those matters that they find more or less important in a given context. It is in light of such values that outcomes are interpreted as positive or negative (see 2.)
4. Social influence (subjective norm)	Intention is also seen to be shaped by how individuals perceive the wishes and norms of relevant others. Social influence is seen to arise from two sub-variables, one relating to norms and the other to motivation (see 5. and 6.)
5. Normative beliefs	This sub-component of social influence relates to the perceptions that individuals have with regard to how relevant others would evaluate adoption or non-adoption. Would they approve or disapprove if they would perform specific behaviour?
6. Motivation to comply	This sub-component of social influence relates to whether or not individuals are inclined to follow what others would like them to do (see 5.). This involves an evaluation of how important and relevant the views of others are
7. Ability (perceived self-efficacy, perceived behavioural control, perceived ease of use, complexity, facilitating factors, effort expectancy, barriers)	Many models include one or more variables that relate to whether or not individuals perceive that performing the behaviour (or adopting the technology) is easy or difficult in light of their own abilities, capacities and self-confidence. Similarly, some models refer to the existence of (real or perceived) external barriers or facilitators that may enable or disable individuals in performing the behaviour. This ability component is mostly regarded as a third influence on intention (see 1.) or as a factor that directly influences adoption behaviour (thus explaining the frequently observed 'intention-behaviour' gap)

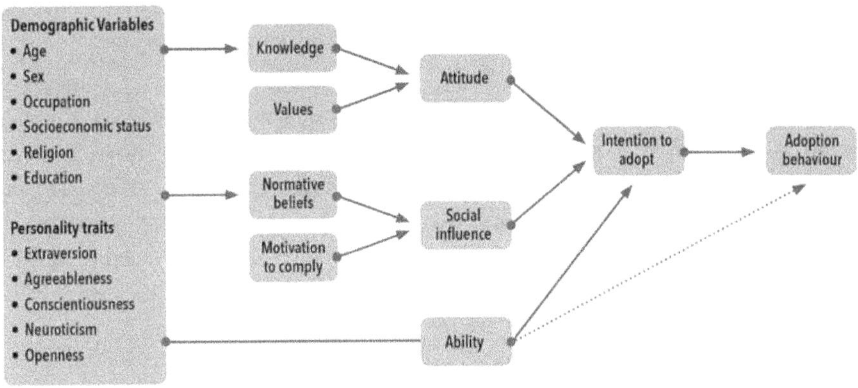

Fig. 4.1 Relationships between main categories of determinants influencing adoption behaviour, as synthesised from various models (see Table 4.2 for related terminology)

credit and/or whether farm labourers will be able to make themselves available at the required time of harvesting. If indeed the variety is new, it is likely to require adapted behaviours of actors along the value chain, which means that adoption of the variety by the farmer is dependent on the simultaneous adoption of complementary behaviours by others in the value chain.

Horizontal interdependencies There are many practices in farming that farmers can only sensibly adopt when other farmers adopt them as well or where adoption is otherwise dependent on the behaviour of other farmers. There exist many crop protection strategies, for example, that are only effective when implemented by the large majority of farmers. If not, sources of infection and disease will continue to exist, rendering adoption useless. Similarly, the growing of new varieties or crops may only be profitable when sufficient farmers in the region join, as buyers may require a certain volume meeting specific quality requirements. Or somewhat differently, farmers in an irrigation scheme or catchment area may only be able to grow a new variety or crop if upstream farmers refrain from using all the water that is available. Again, we see that adoption by one farmer is dependent on the simultaneous adoption of the same or complementary behaviours by other farmers.

Intra-individual interdependencies The adoption of a specific behaviour or technology by an individual often presupposes the acceptance and/or implementation of complementary behaviours by the same individual at more or less the same period in time. Adopting the use of sanitary gloves when milking cows requires several 'preparatory behaviours' such as buying the gloves and instructing family members how to use them. Similarly, adopting a new crop variety may necessitate an earlier or later planting or harvesting date, as well as a different spacing of plants in the field. In other words, in actual practice, 'adoption' involves a plurality of behaviours, and all of these are shaped by the kinds of 'determinants' mentioned in

Table 4.2. This raises the difficult question of precisely which behaviour one should focus on when aiming to understand why people adopt something or not.

Temporal interdependencies Whether or not people are interested to adopt a behaviour or technology may well depend on adoption decisions that have occurred in the past or on decisions that are anticipated in the future. A farmer who has made considerable investments in chemical spraying equipment may not be interested in switching to biological pest control. Similarly, someone who is anticipating to shift from milk to meat production in the long run may not be prepared to consider new approaches to enhance animal welfare for dairy cattle.

Clearly, in view of these interdependencies, the adoption of a particular behaviour by an individual cannot be understood in isolation from a plurality of behaviours and decisions in the broader environment and context of that person (Leeuwis 2004; Van der Ploeg 1990). Therefore, we propose that adoption must be regarded as a relational issue – as something that takes place within a network of relationships between people in time and space. Since considerations related to interdependencies and relationships are not well represented in the individualist and somewhat rationalist perspective on adoption, we will now explore relevant factors and variables that need to be included in an alternative model for explaining reasons for (non)adoption.

4.3.2 Adding Complementary Variables

Issues pertaining to social relations and interdependency may enter into an individual's reasoning in various ways. Below we translate these into several additional perceptual variables.

Trust The insight that people are faced with vertical and horizontal interdependencies when considering adoption can in part be captured by the notion of trust (De Vries et al. 2015), more specifically the expectation that others can be trusted to perform the necessary complementary behaviours and/or provide conducive conditions that enable the adoption of a technology or behaviour. Thus, this variable has connotations of perceived collective ability or perceived environmental effectiveness (Leeuwis 2004). In the context of agriculture, this may relate to whether actors in the agro-support environment or value chain are expected to behave in a conducive manner ('vertical trust') or whether colleagues, household and community members will demonstrate the complementary behaviours on which adoption depends ('horizontal trust'). Another type of trust that may play a role relates to whether the people who are seen to promote the adoption of something are seen to be trustworthy in terms of their expertise, honesty, credibility and legitimacy ('trust in the source').

Responsibility In the context of social relationships, people may well consider whether indeed they are and/or should be responsible for addressing the issue at hand. There are many issues in agriculture where such considerations play a role. In the management of pest and diseases, for example, farmers may refuse to adopt certain pest control practices because they regard it as a government responsibility to organise spraying gangs (Dormon 2006). Similarly, farmers may not want to adapt their farming in a manner that saves the environment because they find that others should take their responsibility first.

Social pressure While the dominant social-psychological models typically include a variable related to social influence, the issue of power and enforcement is not very explicit. From a relational perspective, it is important to acknowledge that some people may be seen to have power over others and can impose sanctions or provide rewards to ensure that others behave in a particular way. For example, female farmers may be discouraged from adopting a cash crop because their husbands forbid it and threaten them to reduce their access to land. Alternatively, people may adopt a new variety purely because they expect to face negative consequences from their community leaders if they refuse. Thus, social pressure is a form of social influence that involves dynamics of power, which make it meaningfully different from how individuals perceive the wishes and norms of relevant others (as considered in social-psychological models).

Trade-offs The observation that adoption involves multiple practices and activities further clarifies that adoption may be considered vis-a-vis a range of goals and aspirations. Clearly, people aim at realising a range of goals in life (including in farming). Hereby, different activities and behaviours may serve different goals, but it is also clear that one and the same behaviour may have implications for different values and aspirations. Buying a tractor for ploughing may be economically attractive but may also damage relations with neighbours and/or farm labourers who were involved in ploughing before the arrival of mechanical ploughing. While social-psychological models involve value-based evaluation of a specific behaviour, it is less clear and explicit that adoption normally involves many behaviours and multiple values at the same time, leading to the likelihood of trade-offs, that is, situations where adoption works out positively for some goals and negatively for others, resulting in the need to make choices and accept certain trade-offs. Different types of goals and aspirations may play a role in the consideration of trade-offs, including economic goals (e.g. high profit or stable cash flow), technical goals (e.g. high milk yield or low water run-off), cultural goals (e.g. showing respect for life), relational goals (e.g. maintaining good relations with neighbours), emotional goals (e.g. safeguarding peace of mind) or ethical aspirations (e.g. ensuring fair and equitable outcomes). The balancing and weighing of multiple perceived outcomes and goal orientations deserve to be made more explicit in models to explain adoption.

Identity It is important to consider that the salient identity that people have or take on when considering something may differ according to the relational setting.

People tend to have several roles and identities (e.g. be a consumer, a producer, entrepreneur, a father, a husband, a citizen), and they may reason and think differently in different capacities. For example, as a citizen, someone may be all in favour of sustainable agriculture and vote for a green party that proposes a tax on meat or fertiliser, but as a consumer, the same person may have different considerations and buy the cheapest meat available, while as a parent he/she may insist on buying the most expensive meat when their grown-up children come home for dinner. From the perspective of an outsider, such behaviours may be regarded as 'inconsistent', but at the same time, one could argue that they are consistent with the salient identity that is evoked by a certain interaction setting. Thus, the salient identity can be seen to shape the kinds of knowledge, values and social influences that are taken into account when deciding to adopt something or not. Arguably, identity can be important in another way as well, namely, an important motivator for action in specific situations. When people feel that their identity is threatened or treated disrespectfully (e.g. their religious, national or racial identity), they may respond strongly. In this sense too, identity may play a role in adoption of technologies or behaviours. For example, people may resist re-allocation of land in the context of a land consolidation or resettlement scheme because they feel it undermines their identity as landowners who have been custodians of the land for many generations.

Risk and uncertainty The existence of numerous interdependencies in relation to adoption only amplifies and underscores that people may well experience uncertainties or risks in relation to several of the variables discussed so far. They may actually wonder what the consequences of one practice may imply for other practices and outcomes or have doubts about how relevant others might respond to their adoption behaviour. Clearly, people are not all knowing about the world around them and may neither be fully convinced about the adequacy of their beliefs, the level of their capacities and/or the importance of competing aspirations. It is likely that such uncertainties and/or risk perceptions have an impact on people who actively consider the adoption of something, but this is not well represented in the dominant models.

Institutions While the complementary variables mentioned so far relate to people's mindsets and mental considerations, it is useful to introduce another type of concept from a rather different academic tradition: the idea that practices or behaviours are shaped by 'institutions'. Institutions can be seen as the formal and inform rules and arrangements to which people orient themselves in their (inter)action (North 1990). We may think of the regulations that govern interactions in markets and value chains, the informal standards and norms regarding appropriate social behaviour, the formal legal rules that people take into account when deciding how to deal with a conflict and/or the procedures that are in place to arrive at democratic decision-making. Clearly, such rules and arrangements (in some literatures called 'structures' or 'regimes') can have an important influence on what people do or do not do. As mentioned, institutional explanations of behaviour stem from a completely different tradition than social-psychological and individualist explanations but certainly merit attention when discussing interdependencies in the context of adoption. In fact, we see that more recent perspectives on innovation than the traditional view described by Rogers (1995) pay considerable attention to issues pertaining to institutions. Currently dominant institutional configurations (or regimes) are often regarded as an obstacle to change and innovation (Geels and Schot 2007); for example, prevailing flat price setting arrangements in the value chain may obstruct the use of quality-enhancing technologies by farmers. At the same time, the existence of such obstacles leads to a call for 'institutional innovation' as an integral component of socio-technical transformation, in this example, the introduction of quality control and price differentiation arrangements to enable farmers to invest in quality-enhancing practices and technologies (Leeuwis 2013). Moreover, innovation processes themselves are seen to be influenced by institutional set-ups that govern interaction between actors in an 'innovation system' or 'innovation ecology' (Hall et al. 2007; Leeuwis et al. 2017), for example, the reward systems in universities may hinder conducive interaction between researchers from different disciplines and/or prevent scholar to engage actively with societal stakeholders.

4.4 Linking Institutional Explanations to Individual Explanations

The question then is how institutional explanations and more individual considerations relate to each other and how the former may be taken into account in a model that complements (but also still builds on) an individualist perspective. We would propose that institutions tend to create resources and (dis)incentives that may enter people's considerations in several ways.

Clearly, rules and arrangements can be seen as shaping the ways in which social influence and social pressure may be exerted or experienced. For example, the standards and rules implied in certification and pricing systems may be used by traders as a leverage to negotiate prices and/or provide (or deny) access to certain market channels and thus operate as an (dis)incentive for farmers to produce in a certain manner. At the same time, such (dis)incentives are likely to affect people's aspirations and alter the balance in assessing trade-offs between different goals. Moreover, the existence of community by-laws and accompanying control and sanctioning systems may affect the extent to which farmers trust their neighbours in adhering to preventive measures against diseases. Institutions may also shape the knowledge and beliefs that people have about the world and/or about the outcomes of specific behaviours. The belonging to certain scientific disciplines, religious groups or cultural communities, for example, impinges on how people consider their relation with nature (Douglas 1970) and also on agricultural knowledge and beliefs. Organic farmers, for example, tend to have a very different belief system regarding the control of diseases and/or the management of soil fertility than so-called 'regular' farmers. Similarly, such cultural belongings and identities may relate to what people feel responsible for and what not.

Thus, in a setting where multiple people interact and depend on each other, institutions can be seen to orient actors' ways of thinking, that is, their beliefs, experienced influence or pressure, aspirations, trust, risk perceptions, abilities, responsibilities, etc. This should not be interpreted in a deterministic manner as actors can have considerable agency and space for manoeuvre (Long 1990; Giddens 1984) to interpret what rules and arrangements apply to the situation at hand.

Figure 4.2 captures what we have discussed so far:

Figure 4.2 provides what one could call a 'sociologically enhanced' overview of the reasons that people may have for adopting or rejecting an innovation (consisting of a package of behaviours A, B, C). When compared with Fig. 4.1, it includes a range of additional relational variables and considerations that have to do with how other people are looked at: whether they exert power and pressure, whether they can be trusted to (be willing and able to) perform complementary behaviours and/or whether proposed responsibilities and identities are meaningful to the situation. The figure also captures the role of salient identities and institutions as well as the role of time; it makes clear that previous behaviours and experiences play a role and that the adoption of a set of new practices generates responses in the agro-ecological and social world that are being interpreted, leading to changes in people's 'mindset'. For example, growing a new variety may go along with a variety of experiences which are likely to influence peoples' reasoning with regard to future activities: Did it grow well?, Was it resistant to diseases?, How much effort was needed?, Was I able to sell the crop against a good price?, Did my family members like the taste?, etc. Once behaviours associated with an innovation have been adopted (and often adapted), they are likely to become a routine or habit, that is, a regular practice that is no longer actively deliberated every time it is enacted. This is why the concept of 'intention' has been omitted from Fig. 4.2; while behaviours may continue to be rationalised ex-post (e.g. when asked about in an interview setting), it is not accurate to assume that everything people do is preceded by careful deliberation. However, at some point in time, routines may become subject to active reflection again, for example, when a relatively stable situation is disrupted and/or when longer-term negative consequences become visible.

4.5 An Interactional View

Whereas Fig. 4.2 still gravitates towards an individualist perspective as it includes only one relatively isolated 'mindset', Fig. 4.3 is an attempt to correct that to some degree.

While Fig. 4.3 refers to the same parameters as Fig. 4.2, it highlights that adoption occurs in an interactional context where several people (in reality more than two) depend on each other in their performance. This interaction evokes salient identities of participants and is oriented by the formal and informal institutions that

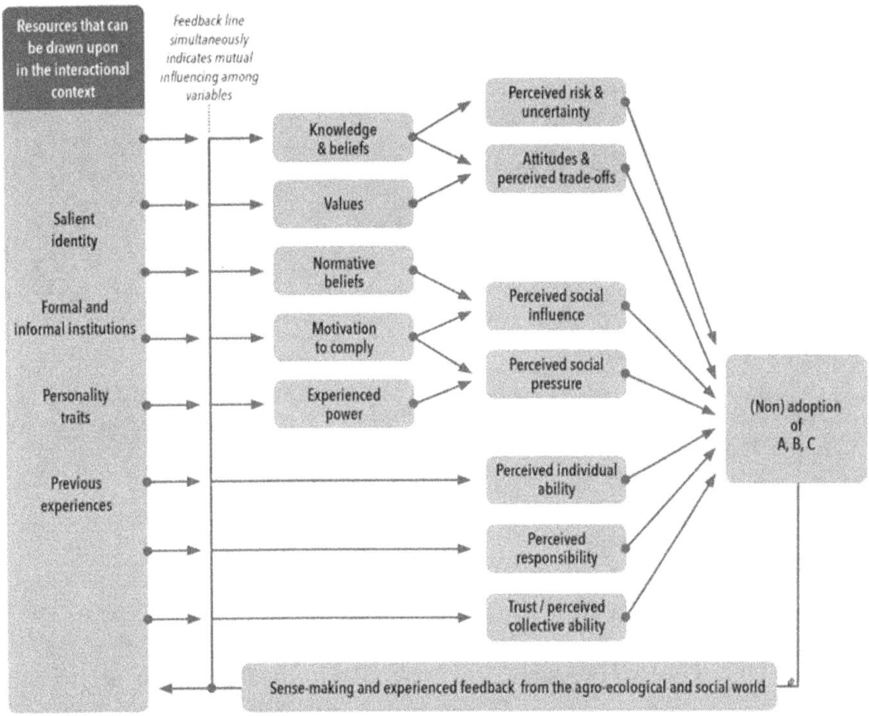

Fig. 4.2 Reasons for (non)adoption: complementing individual determinants with variables arising from interdependence and institutional influences

actors draw upon. The 'mindsets' of the actors (their 'reasons for (non)adoption', see Fig. 4.2) are influenced by the exchange that takes place, by the elements in the institutional setting that are deemed relevant as well as by contextual conditions, personality traits and previous experiences. At any point in time, the participants are likely to have meaningfully different 'mindsets' and/or priorities in their reasoning, which are bound to change and evolve during the process. As an outcome of the interaction, the people involved may or may not adopt complementary practices (A, B, C and X, Y, Z), or enable or constrain each other to move in a given direction.

The image makes clear that in order to move towards the adoption of a specific package of interrelated practices (say prevention of diseases through integrated pest management practices in a community), the actors involved must somehow develop 'mindsets' that are congruent with each other. This may be in the form of jointly agreed upon rules or by-laws (institutions), overlapping aspirations (values), shared understandings of what will happen if A, B, C and X, Y, Z are performed (knowledge and beliefs) and mutual recognition of the rewards and sanctions that will be applied (social pressure). Arriving at such mutual adjustment and understanding involves different (or at least additional) processes than the individual decision-making processes portrayed in Table 4.1. As argued and elaborated elsewhere

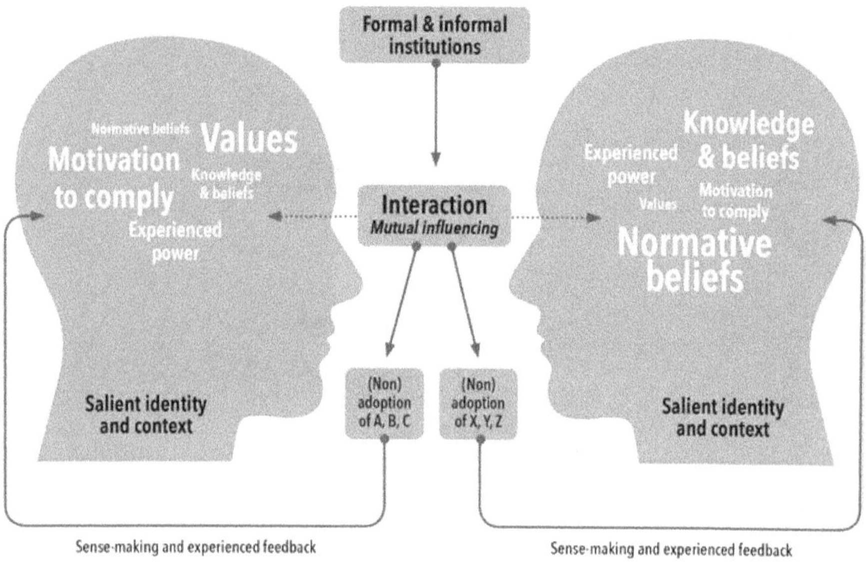

Fig. 4.3 An interactional view of reasons for (non)adoption

(Leeuwis 2004; Leeuwis and Aarts 2011; Aarts 2018b), mutual adjustment typically involves processes of social learning, negotiation and conflict resolution. Such processes may be facilitated by 'innovation intermediaries' that play broader roles than providing individual advisory services and/or applying persuasive strategies geared towards changing individual behaviour (see for details Leeuwis 2004; Klerkx and Leeuwis 2008; Klerkx et al. 2009).

4.6 Implications for Development Practice: Rethinking Scaling and Information Provision Through ICT4Ag

As indicated at the start, fostering agricultural development in one direction or another always involves changes in human practices and hence the 'adoption' of new technological and/or social-organisational behaviours. From the perspective of development practitioners, it is relevant to ask what the practical implications are of our 'interactional' and 'sociologically enhanced' perspective on 'reasons for (non) adoption'. Below, we will apply our perspective to two topical issues that actors in the field of agricultural development tend to struggle with in different contexts. The first issue involves the idea of achieving development impact through the 'scaling' of innovations, and the second issue relates to information provision through 'ICT4Ag' (information and communication technology for agriculture) as a means of supporting processes of adoption and scaling.

4.6.1 Sharpening Our Thinking About 'Scaling'

While the term 'scaling' has some broader connotations (Wigboldus 2018; Wigboldus et al. 2016), it is often used in a way that resembles the older terminology of 'adoption and diffusion of innovations' as used by Rogers (1962). Both terminologies often express a normative desire to ensure that something that is considered to be good and desirable spreads across a greater number of users and/or across a larger geographical area, in order to achieve some kind of societal impact. 'Scaling' is seen as a critically important process and frequently also as an enormous challenge. Our model of 'reasons for (non)adoption' may help in several ways to sharpen our thinking about 'scaling'.

The plurality of scaling: thinking in terms of assemblages The 'reasons for (non) adoption' model is a response to the existence of several types of interdependencies in a development setting (see above). These interdependencies imply that one cannot usefully consider the scaling of one particular practice in isolation from other practices, including practices that are performed by other people than the originally perceived 'users'. Similarly, the 'upscaling' of one practice is likely to require the 'downscaling' of other practices that are being replaced or affected. Hence, it is important to think about scaling in terms of multiple practices in an assemblage that are simultaneously scaling up or down (Leeuwis and Wigboldus 2017; Sartas et al. 2019). Identifying the interdependent practices in a network of interdependent stakeholders (and in time) provides useful insights in the complexity of a particular scaling ambition (see Fig. 4.4 and Sartas et al. 2019; Sartas et al. 2020).

Targeting 'bottleneck' or 'leverage' practices as entry points Having insight in an assemblage of interdependent practices begs the critically important question of which practice(s) could or should be the focus of attention in a development effort (Sartas et al. 2019). While the original entry point might be the wish to 'scale' a drought-resistant variety, a consideration of the broader assemblage of practices involved may well lead to the identification of more relevant entry points for intervention. The use of the variety by farmers may, for example, be constrained by prevailing policies that prevent the release or distribution of the variety, thus shifting the attention to interventions that may help to overcome such policy bottlenecks rather than simply promote the variety. Alternatively, one may conclude that the use of the variety can be leveraged by the creation of specific price incentives offered by a dominant trader or wholesaler. In a complex environment of interrelated practices, it is an illusion to think that one can bring about change by focussing on one particular issue or practice, while at the same time, it is unrealistic and inefficient to target interventions on everything that matters. Thus, interventionists need to somehow identify the critical interdependencies and leverages in the system in order to develop an effective scaling strategy (Sartas et al. 2019; Vellema and Leeuwis 2019).

Diagnosing with the help of the model Finding the leverage or bottleneck practices requires an understanding of stakeholders' (interactional) rationale in relation

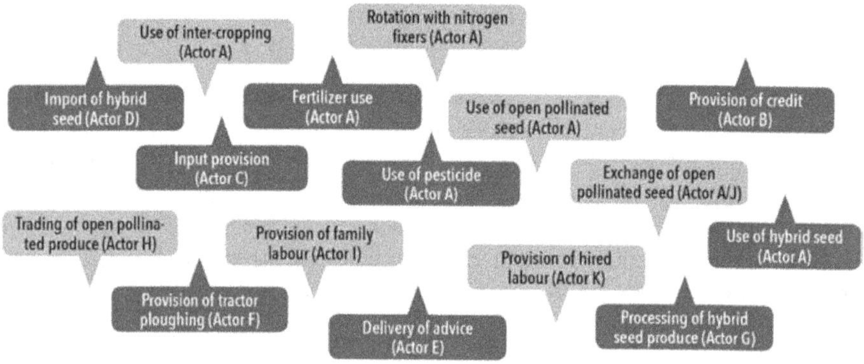

Fig. 4.4 Fictitious example of multiple practices scaling 'up' and 'down' simultaneously in an assemblage

to existing and/or alternative practices. In other words, what are the reasons that underpin current practices and interaction patterns and/or what are the reasons that actors have for rejecting alternative practices and courses of action? The 'reasons for (non)adoption' model (Figs. 4.2 and 4.3) offers several entry points of investigating and diagnosing these.

First, the model offers a more elaborate overview of relevant variables than social-psychological models, which can be used to describe and analyse people's rationale. For example, if one wants to understand why farmers refuse to spray against a disease, one may consider how this relates to their knowledge (how do they understand the disease dynamics?), their feelings of responsibility (do they feel responsible for combating the disease?), their values (how important is the crop for them?), the social influence or social pressures experienced (what do others expect and what sanctions or incentives are in place?), their trust in others (are farmers confident that others will spray as well and/or that the agro-dealer sells the right chemicals?) and their individual abilities (are farmers confident that they have the right skills and equipment for spraying and/or can they afford to obtain these?). Disentangling people's rationale in relation to several practices in an assemblage helps to understand what the most important bottlenecks or leverages may be. Simultaneously, this offers entry points for intervention: if farmers refuse to spray because they do not understand the dynamics of the disease, then it makes sense to invest in interventions that foster awareness raising and learning. However, if farmers do not feel responsible or do not trust that their neighbours will also spray, then other types of strategies will have to be considered.

A second entry point for diagnostic analyses offered by the model is to focus on the formal and informal institutions that orient people's considerations and ways of thinking. When the purpose is to develop a scaling strategy, such analysis of people's rationales and underlying institutions needs to be combined with an assessment of which critically bottlenecks and leverages may be(come) amenable to change and what types of interventions (ranging from persuasive campaigns to participatory design in a multi-stakeholder process) may be conducive to achieving this (Sartas et al. 2019). As already hinted at, the realisation that scaling involves many interdependencies and a plurality of practices and stakeholders implies that approaches that allow for multi-stakeholder learning and negotiation are likely to be relevant in many instances (Leeuwis and Aarts 2011).

Responsible scaling Thinking in terms of interdependent practices between stakeholders also highlights that scaling a particular practice may trigger and/or result in the scaling of other practices and phenomena that were not initially considered. Thus, new practices may yield positive or negative consequences for those directly involved and also generate outcomes that occur at other levels of aggregation. The use of a new drought-resistant variety may, for example, make households dependent on credit for buying seeds, which – in case of high interest rates – may negatively affect households' resilience or ability to pay school fees. Similarly, the application of new oil palm production systems at a large scale may have negative consequences for biodiversity in a region, or put farmers in another region out of business. Investigating and anticipating these kinds of interrelations is part and parcel of a broader approach that Wigboldus has labelled 'responsible scaling' (Wigboldus 2018) – the careful consideration of possible positive and negative consequences of scaling with regard to diverse societal values and categories of people, as well as principles of ethics and democracy. Thinking in terms of interdependencies between practices is a useful way of starting such a process (see for more elaboration and guidance Wigboldus and Brouwers 2016).

The kind of thinking presented here about scaling has been further translated into tools and methods that may support the development of scaling strategies elsewhere (Sartas et al. 2019).

4.6.2 Rethinking Information Provision Through ICT4Ag: The Example of Disease Control

The early work of Rogers (1962) and Van den Ban (1963) already highlighted the importance of communication and information provision (see Table 1) in supporting adoption and scaling processes. In the current age of enhanced mobile phone and internet connectivity in developing countries (Dey et al. 2016; De Bruijn and Van Dijk 2012), there is a lot of attention to how digital ICT platforms may be leveraged to enhance scaling. Therefore, a relevant question is how our 'interactional' and 'sociologically enhanced' understanding of 'reasons for (non)adoption' may impinge on communicative intervention and information provision through ICT (or other media).

First of all, it is relevant to note that the individualist and rationalist perspective on adoption has greatly inspired and influenced communicative interventions geared towards agricultural development, as exemplified in the practice of agricultural extension (Van den Ban and Hawkins 1996). Typically, agricultural extension handbooks and extension professionals identify strongly with the idea that individuals need to be provided with relevant information that guides them through the adoption process, that is, provision of information about problems, solutions, pros and cons associated with alternative options, etc. geared largely towards changing people's beliefs and attitudes (see Table 4.2) in favour of specific behaviours. Classically, such information was communicated through a mix of mass media and interpersonal media. While the media may have changed, it is interesting to note that ICT applications in agricultural extension may still tend towards provision of similar kinds of information to farmers. Recent inventories of ICT4Ag indicate that extension organisations use virtual platforms to enhance organisational processes through better registration of farmers, recording of activities and internal reporting and also to provide farmers with up-to-date weather and market information (Munthali et al. 2018). When it comes to technical advice that is explicitly geared towards influencing adoption, we see that ICTs are frequently used to provide information about 'best practices' through, e.g. repositories, voice messages, text messages, video clips and alerts. While the channel and speed through which information is provided may have meaningfully changed, such information arguably is still directed at individual farmers and contains similar messages as in traditional extension regarding the existence of problems and solutions and the pros and cons of different courses of action.

If indeed ICT4Ag continues to be geared towards influencing beliefs, attitudes and individual adoption decisions, it is likely to fall short in similar ways as pre-existing forms of decision support in that these do not directly address interdependencies between people and practices. Thus, it is worthwhile to think about what ICT4Ag might have to offer in terms of supporting processes of collective decision-making on farmer-level agricultural issues. Phrased differently, how may ICT4Ag help to anticipate influential interactional processes, and be used to shape the more relational variables in our 'reasons for (non)adoption' model, including social pressure, power, trust, responsibility, salient identity and institutions?

To explore this further, it may help to think of an important farmer-level issue in which interdependencies among farmers tend to play an important role: the prevention and control of pests and diseases. There exist numerous plant diseases whereby the efficacy of control measures on one farm depends on what other farmers in the vicinity do. If farmers in the immediate environment do not take sufficient preventive or curative measures (e.g. disinfect tools, prevent water run-off, remove and burn diseased plants, apply spraying at the right time, install insect traps, buy clean planting materials, etc.), it becomes almost futile for a farmer to invest in disease control on his or her own, since the field will continue to become infected by the disease. In such cases, diseases can be seen as a 'public bad', while effective disease control strategies can be regarded as a 'public good' that is only created if sufficient farmers contribute to it (Cieslik et al. 2018; Leeuwis et al. 2018). In these kinds of situations, it is clearly insufficient to only provide individuals with technical advise on how to prevent and control the disease; even if farmers come to belief that such measures are likely to be effective and develop a positive attitude towards them, they are unlikely to perform them unless they are reasonably sure that their neighbours will take proper action as well.

In connection with these kinds of situations, Ostrom (1990, 2009) has identified several communicative and informational conditions and strategies related to the interaction between interdependent actors that are conducive to creating a 'public good' (e.g. an effective community-based disease management strategy). Below, we briefly discuss these conditions and how they link to our 'reasons for (non)adoption' model and the possibilities of ICT4Ag.

Typically, the effective maintenance of a common pool resource and/or the creation of a public good requires the existence of certain rules (institutions) with regard to how people in a community of actors should behave. In relation to pest and disease management, such rules could be 'to remove and burn diseased plants as soon as possible', to 'build ditches around diseased fields to prevent infection through run-off water', to 'apply preventive spraying after the first rains have passed' and/or 'to make a monthly contribution to cover maintenance costs of collective spraying equipment'. Ostrom (1990, 2009) concludes that in order for such rules to be effective, it is important that most individuals in the community are able to participate in making and modifying them. Clearly, this requires intensive communication between interdependent actors in the community. Similarly, fostering adherence to such rules depends on the availability of various kinds of information. According to Ostrom, members of the community need an up-to-date information about the condition of the resource that is relevant and actionable in view of the prevailing rules, in this case, information about the agro-ecological conditions of the field and the actual presence of the disease. In addition, community members are more likely to conform to the rules if they have information about the behaviour of others on which they depend, for example, information with regard to whether or not others are fulfilling their obligations or not and whether or not sanctioning systems operate effectively. Clearly, the generation and distribution of such information requires the operation of a monitoring system that captures both agro-ecological conditions and human behaviour and makes them available to those belonging to the community.

Clearly, the kinds of communicative and informational functions mentioned above can potentially be supported by ICT. Social media applications may, for example, support interaction within a community of actors during the process of designing rules and even help enlarge the boundaries of effective community formation and identity building (Cieslik et al. 2018; Bennett and Segerberg 2012). Similarly, mobile phones can serve to record, report and process decentralised observations as part of a community-based monitoring system for pest and disease management and help to share such information with participating farmers. In this manner, communicative and informational services may help to foster conducive conditions for collective action in response to agricultural pests and diseases. It must be noted that such types of ICT applications would differ markedly from those oriented towards disseminating 'best practices' and/or persuading individual farmers to adopt them. Rather than focussing on influencing 'determinants' for individual behaviour (e.g. knowledge, attitudes, ability), they are geared towards supporting collective identity formation, design of institutions, maintenance of trust and the effective use of power and sanctioning systems in a community of actors. Thus, our 'interactional' and 'sociologically enhanced' understanding of 'reasons for (non) adoption' helps us to imagine different kinds of ICT4Ag applications than those that just provide regular extension services through a different medium.

4.7 Concluding Remarks

While recent critiques of the notion of adoption emphasise the shortcomings of the technocentric and binary thinking that is often implied, this chapter has focussed on the limitations in dominant thinking about the process side of adoption and has broadened the scope by providing an interactional perspective. For a long time, adoption has been portrayed as a largely individual process that could be understood and influenced with the help of social-psychological models. We have argued that this kind of thinking ignores the existence and importance of vertical, horizontal, intra-individual and time-related interdependencies between practices, leading to the conclusion that in many instances, adoption must be regarded as a collective rather than an individual process. We have indicated how interdependencies may enter into an individual's reasoning about adoption in various ways and have translated this into several additional variables that need to be considered when the aim is to explain adoption and behaviour change or the lack of it. Consideration of these variables has led us to develop an 'interactional' and 'sociologically enhanced' model for understanding 'reasons for (non)adoption'. Subsequently, we have explored the practical implications of this mode of thinking by linking them to topical issues such as 'scaling' and the use 'ICT4Ag' in stimulating adoption. This exploration reveals that we can sharpen our thinking about scaling considerably by realising that scaling always involves an assemblage of different interdependent practices across a network of stakeholders in time and space, which necessitates critical thinking about points of leverage and issues of responsibility. Similarly, our exploration of the role that ICT4AG may play in the management of agricultural diseases has revealed that the enhanced model for understanding reasons for (non) adoption solicits the design of novel kinds of ICT applications and services that may support the creation of conducive conditions for collective action. Thus, moving beyond individualist conceptualisations of adoption not only represents a theoretical advance but also helps us to re-imagine and re-orient the kinds of interventions needed to shape adoption processes in support of realising a desired future.

Acknowledgements This research was undertaken as part of, and funded by, the CGIAR Research Program on Roots, Tubers and Bananas (RTB) and supported by CGIAR Trust Fund contributors as well as by NWO and Wageningen University.

References

Aarts MNC (2018a) Dynamics and dependence in socio-ecological interactions. Inaugural address. Radboud University Nijmegen, Nijmegen

Aarts MNC (2018b) Boundary spanning for strategic communication: towards an interactional and dynamic perspective. In: Heath RL, Johansen W (eds). The international encyclopedia of strategic communication, Boston. Wiley – Blackwell Publishers, pp 91–96

Ajzen I (1985) From intentions to actions: a theory of planned behavior. In: Kuhl J, Beckmann J (eds) Action control: from cognition to behavior. Springer, New York, pp 11–39

Ajzen I (1991) The theory of planned behavior. Organization behavior and human decision processes. pp 179–211

Baumeister R, Vohs K, Tice D (2007) The strength model of self-control. Curr Dir Psychol Sci 16(6):351–355

Bennett WL, Segerberg A (2012) The logic of connective action. Inf Commun Soc 15(5):739–768

Cieslik KJ, Leeuwis C, Dewulf ARPJ, Lie R, Werners SE, van Wessel M, Feindt P, Struik PC (2018) Addressing socio-ecological development challenges in the digital age : exploring the potential of environmental virtual observatories for connective action (EVOCA). NJAS Wageningen J Life Sci 86-87:2–11

Davis FD, Bogozzi R, Warshaw PR (1989) User acceptance of computer technology: a comparison of two theoretical models. Manag Sci 35:982–1003

De Bruijn ME, Van Dijk R (2012) Connecting and change in African societies: examples of 'ethnographies of linking' in anthropology. Anthropologica 54:45–59

De Vries JR, Aarts N, Lokhorst AM, Beunen R, Oude Munnink J (2015) Trust-related dynamics in contested land use. A longitudinal study towards trust and distrust in intergroup conflicts in the Baviaanskloof, South Africa. J Forest Policy Econ 50:302–310

Dey B, Sorour K, Filieri R (2016) ICTs in developing countries: research, practices and policy implications. Palgrave Macmillan, New York and London

Dormon ENA (2006) From a technology focus to innovation development : the management of cocoa pests and diseases in Ghana. PhD dissertation, Wageningen University

Douglas M (1970) Natural symbols. Explorations in cosmology. Barrie and Rockliff/Cresset Press, London

Fishbein M, Ajzen I (1975) Belief, attitude, intention, and behavior: an introduction to theory and research. Addison-Wesley Pub. Co., Reading/Don Mills

Geels FW, Schot JW (2007) Typology of sociotechnical transition pathways. Res Policy 36(3):399–417

Giddens A (1984) The constitution of society: outline of the theory of structuration. Polity Press, Cambridge

Glover D, Sumberg J, Andersson JA (2016) The adoption problem; or why we still understand so little about technological change in African agriculture. Outlook Agric 45(1):3–6

Hall A, Clark N, Naik G (2007) Institutional change and innovation capacity: contrasting experiences of promoting small scale irrigation technology in South Asia. Int J Technol Manag Sustain Dev 6(2):77–101

Kahneman D (2011) Thinking, fast and slow. Penguin Books, London

Klerkx L, Leeuwis C (2008) Matching demand and supply in the agricultural knowledge infrastructure: experiences with innovation intermediaries. Food Policy 33(3):260–276

Klerkx LWA, Hall A, Leeuwis C (2009) Strengthening agricultural innovation capacity: are innovation brokers the answer? Int J Agric Res Gov Ecology 8(5–6):409–438

Leeuwis C (with contributions by A. Van den Ban) (2004). Communication for rural innovation. Rethinking agricultural extension. Blackwell Science/CTA, Oxford/Wageningen, 412 p

Leeuwis C (2013) Coupled performance and change-in-the-making. Inaugural lecture. Wageningen University, Wageningen

Leeuwis C, Aarts N (2011) Rethinking communication in innovation processes: creating space for change in complex systems. J Agric Educ Extension 17(1):21–36

Leeuwis C, Wigboldus S (2017) What kinds of 'systems' are we dealing with? Implications for systems research and scaling. In: Öborn I, Vanlauwe B, Phillips M, Thomas R, Brooijmans W, Atta-Krah K (eds) Sustainable intensification in smallholder agriculture. an integrated systems research approach., Earthscan food and agriculture series. London and New York. Routledge Taylor & Francis Group, pp 319–333

Leeuwis C, Schut M, Klerkx L (2017) Systems research in the CGIAR as an arena of struggle:

competing discourses on the embedding of research in development. In: Sumberg J, Andersson J, Thompson J (eds) Agronomy for development: the politics of knowledge in agricultural research. Routledge, pp 59–78

Leeuwis C, Cieslik KJ, Aarts MNC, Dewulf ARPJ, Ludwig F, Werners SE, Struik PC (2018) Reflections on the potential of virtual citizen science platforms to address collective action challenges : lessons and implications for future research. NJAS Wageningen J Life Sci 86-87:146–157

Loevinsohn M, Sumberg J, Diagne A (2012) Under what circumstances and conditions does adoption of technology result in increased agricultural productivity? Protocol, EPPI Centre, Social Science Research Unit, Institute of Education, University of London, London

Loevinsohn M, Sumberg M, Diagne A, Whitfield S (2013) Under what circumstances and conditions does adoption of technology result in increased agricultural productivity? A systematic review. IDS, Brighton

Long N (1990) From paradigm lost to paradigm regained? The case for an actor-oriented sociology of development. Eur Rev Latin Am Caribbean Stud 49:3–32

Munthali N, Leeuwis C, van Paassen A, Lie R, Asare R, van Lammeren R, Schut M (2018) Innovation intermediation in a digital age : comparing public and private new-ICT platforms for agricultural extension in Ghana. NJAS Wageningen J Life Sci 86-87:64–76

North DC (1990) Institutions, institutional change and economic performance. Cambridge University Press, Cambridge

Ostrom E (1990) Governing the commons: the evolution of institutions for collective action. Cambridge University Press, Cambridge

Ostrom E (2009) Beyond markets and states: polycentric governance of complex economic systems. Prize lecture, December 8, 2009. Workshop in Political Theory and Policy Analysis, Indiana University, Bloomington, IN 47408, and Center for the Study of Institutional Diversity, Arizona State University, Tempe

Petty RE, Cacioppo JT (1986) The elaboration likelihood model of persuasion. In: Berkowitz L (ed) Advances in experimental social psychology, 19. Academic Press, New York, pp 123–205

Rana NP, Dwivedi YK (2015) Citizen's adoption of an e-government system: validating extended social cognitive theory (SCT). Gov Inf Q 32(2):172–181

Rogers EM (1962) Diffusion of innovations, 1st edn. Free Press, New York

Rogers EM (1995) Diffusion of innovations, 4th edn. Free Press, New York

Sartas M, Leeuwis C, van Schagen B, Velasco C, Thiele G, Proietti C, Schut M (2019) Scaling readiness quick guide. Concepts and Practices of Scaling Readiness. CGIAR Research Program on Roots, Tubers and Bananas (RTB). July 2019. Available at www.scalingreadiness.org

Sartas M, Schut M, Thiele G, Proietti C, Leeuwis C (2020) Scaling Readiness: Science and practice of an approach to enhance impact or research for development. Agricultural Systems 183

Taherdoost H (2018) A review of technology acceptance and adoption models and theories. Procedia Manufacturing 22:960–967

Van den Ban AW (1963) Boer en landbouwvoorlichting: De communicatie van nieuwe landbouwmethoden. Pudoc, Wageningen

Van den Ban AW, Hawkins HS (1996) Agricultural extension, 2nd edn. Blackwell Science, Oxford

Van der Ploeg JD (1990) Labor, markets, and agricultural production. Westview Press, Boulder

Vellema S, Leeuwis C (2019) Contextualising scaling. Shifting from individual choices to generative processes in networks. Discussion note prepared for the CCAFS NWO-GCP4 Mid-Term Workshop, held in Addis Ababa, June 12–14, 2019

Venkatesh V, Davis FD (2000) A theoretical extension of the technology acceptance model: four longitudinal field studies. Manag Sci 46(2):186–204

Venkatesh V, Morris MG, Davis FD, Davis GB (2003) User acceptance of information technology: toward a unified view. MIS Q 27:425–478

West RL, Turner LH (2010) Uses and gratifications theory. In: West RL, Turner LH (eds) Introducing communication theory: analysis and application. McGraw-Hill, Boston, pp 392–409

Wigboldus S (2018) To scale, or not to scale – that is not the only question: rethinking the idea and practice of scaling innovations for development and progress. PhD dissertation, Wageningen University, London and New York

Wigboldus S, with Brouwers J (2016). Using a theory of scaling to guide decision making. Towards a structured approach to support responsible scaling of innovations in the context of agrifood systems. Wageningen University and Research, Wageningen

Wigboldus S, Klerkx L, Leeuwis C, Schut M, Muilerman S, Jochemsen H (2016) Systemic perspectives on scaling agricultural innovations. A review. Agron Sustain Dev 36

5

Sweden's Agri-Industry: Innovation and the Development of a Sustainable Business Model

Per-Ola Ulvenblad

5.1 Introduction

This chapter focuses on the development of sustainable business models for innovation in the Swedish agri-sector. This is important for several reasons. At the global level, many of society's challenges are linked to social, environmental and economic aspects of agriculture. Worldwide food production must increase by 70% from 2009 to 2050, and in developing countries, the increase needs to be 100% (FAO 2011). However, productivity growth has fallen and remains below potential in many countries (OECD 2019). Simultaneously, the negative climate impact of agriculture has to be reduced. The recently released IPCC Special Report on Climate Change and Land highlights the interconnection between climate change and the agri-sector (including forestry). At the same time, as some parts of the agri-sector are drivers of climate change, sustainable agriculture and forestry can reduce climate change (IPCC 2019). Further, research has suggested that investment in agriculture is an effective strategy to achieve many of society's development goals, like poverty and hunger, nutrition and health, education, economic and social growth, peace and security and preserving the world's environment (Dobermann and Nelson 2013).

The IPCC report also states that climate change represents a threat to the agri-sector (IPCC 2019). As research has identified, there is a need to focus R&D on how to improve agricultural adaptation to climate change, especially in food-insecure human populations (e.g. Lobell et al. 2008). Furthermore, sustainable food production will have to address aspects such as the nutritional quality of diets and positive health effects (Benbrook et al. 2013; Średnicka-Tober et al. 2016).

P.-O. Ulvenblad (✉)
School of Business, Innovation and Sustainability, Halmstad University,
Halmstad, Sweden
e-mail: Per-Ola.Ulvenblad@hh.se

Agricultural research has made important contributions to poverty reduction and food security over the last 40 years (Thornton et al. 2017). Research has also identified investment in R&D as an important driver of growth in agricultural production efficiency (Alston 2010, 2018; Fuglie et al. 2017). Regardless, European Union's Research and Innovation Programme "Horizon 2020" (European Commission 2011) and United Nations' "The 2030 Agenda for Sustainable Development" including the 17 sustainable development goals (SDG) (Griggs et al. 2013; United Nations 2015; European Commission 2016) call attention to the need for more research and innovation on food security and sustainable agriculture.

At the firm level, many agri-companies often struggle with low profitability. In order to be able to produce, distribute and sell more food, they need to achieve profit goals. The majority of agri-companies have focused on their role as producer at the beginning of the food value chain and, consequently, have focused on becoming more effective and being able to produce more with the same, or less, resources. However, the issue of profitability remains an issue for agri-companies (Dobermann and Nelson 2013; Ulvenblad et al. 2016), and they face increasing demands from governments, local authorities, other companies in the agri-value chain and end customers regarding quality and sustainability issues.

The focus on efficiency, economies of scale and growth has been an effective approach for the agri-sector (Alston 2018). Over time, many of the small cooperatives and networking firms in the agri-sector have either joined or become large multinational companies through mergers and acquisitions. Though this trend has led to cost-effective production and distribution systems, has it also built barriers for sustainable business model innovation in the agri-value chain?

In recent times, small agri-companies have often been reduced to a subcontractor role without any real influence (Ulvenblad et al. 2016). It is known that power asymmetry combined with low-quality business relationships can lead to suboptimization and a reduced ability to identify and meet end-consumer needs (Benton and Maloni 2005; Schulze-Ehlers et al. 2014). Large companies mainly look for standardized products that suppliers produced at low cost, and innovations developed by smaller companies along the food chain making products better but different can be difficult to integrate into big companies' business plans and delivery systems.

These are important issues, because of the increased worldwide competition, advanced technological developments and large-scale production existing in the agri-sector, and with a resulting general trend towards fewer and larger farms (OECD 2016), it remains to be seen whether the global goals of sustainable agricultural development can be met.

The agri-sector is different from other industries for several reasons: food from living things, animals and plants, must meet specific welfare, health and safety requirements. Furthermore, production and often distribution are generally connected to a specific geographic area, where nature and climate may have important influence and establish constraints to production and distribution. OECD (2019) highlights the need for more responsive agricultural innovation systems, since climate change and weather-related production shocks are expected to increase the challenge of improving productivity, sustainability and resilience on farms.

When increasing productivity, efficiency and economies of scale have become very important in the agri-sector, small- and medium-sized agri-companies engaged in business model innovation focusing on sustainability must take a more strategic and innovative perspective. A strategy for these companies can be to expand their business not through traditional growth, but rather by diversifying activities and focusing on cooperation. Research have shown that small- and medium-sized agri-companies that have been able to overcome the challenges of the agri-sector have developed sustainable business models based on a diversified approach (Ulvenblad et al. 2016), a network approach (Lawson et al. 2008) or a value-net approach (Kähkönen 2012). These sustainable business models often generate value not only for customers but also for other stakeholders, the community and the environment (Barnett and Salomon 2012; Kiron et al. 2013; Schaltegger et al. 2016).

Previous research regarding agri-companies have, in general terms, focused on production and cost-efficiency (Alston 2010, 2018). Further, research regarding business models has often addressed industries other than the agri-sector (Tell et al. 2016) such as media, information technology and biotechnology industries (Johnson 2010).

Schaltegger et al. (2016) state that "The business model perspective is particularly interesting in the context of sustainability because it highlights the value creation logic of an organization and its effects and potentially allows (and calls) for new governance forms such as cooperatives, public private partnerships, or social businesses, thus helping transcend narrow for-profit and profit-maximizing models" (p.5). This is line with Boons et al. (2013), who conclude that business models are useful for the creation and study of sustainable innovation. Lambert and Davidson (2013) and Zott et al. (2011) have shown that a sustainable business model is a useful tool when studying how a single company or an entire value network achieves sustainability in terms of environmental, social as well as economic value. However, the existing research on sustainable business models in the agri-sector has often been limited to and addresses developing rather than developed countries (e.g. the United States and European countries) (Beuchelt and Zeller 2012). Even though a structured literature review reveals an increasing number of research articles that examine sustainable business model innovation in the agri-sector, there is a need for more research on this topic (Tell et al. 2016).

To summarize, businesses in the agri-sector face difficult challenges – but have also unique opportunities to develop sustainability-oriented innovation and sustainable business models which create value in other ways than low-cost production. However, even if the needs at the global and firm levels are large, research regarding the development of sustainable business models within the agri-sector is limited. This is a significant issue, since profitable agri-companies, which develop sustainable business models and advance along the agri-value chain, need to become part of the solution at a moment in history when the world needs more food, which is to be produced and distributed in sustainable ways.

In order to reduce the research gap identified, the aim of this chapter is to illustrate and analyse how Swedish agri-companies strive to develop sustainable business models to innovate their business activities. I will illustrate, map, categorize and analyse the orientation of the practices of sustainability innovation and the development of sustainable business models. The research questions are: (i) "How do Swedish agri-companies apply sustainable innovation practices?", (ii) "Which sustainable business models do Swedish agri-companies use?", and (iii) "How can these innovation practices and sustainable business models be understood?"

The remaining of this chapter is organized as follows: A conceptual framework will be articulated, and, subsequently, the methodological research approach will be presented. Later on, eight Swedish cases including companies and cooperatives will be illustrated and analysed. Once these cases are discussed, some aspects of theory development will be developed. Finally, based on the most salient learnings from the research, guidelines for agri-entrepreneurs will be provided as well as suggestions for future research.

5.2 Conceptual Framework

The theory section is organized as follows: sustainability in the agri-sector, sustainability-oriented innovation, sustainable business models and eight types of sustainable business models.

5.2.1 Sustainability in the Agri-sector

The field of sustainability is a relatively new discipline that has developed over the last 15–20 years (Coad and Pritchard 2017). Nidumolu et al. (2009) showed a decade ago that sustainability became innovation's new frontier. They stated that sustainability is a fundamental layer of organizational and technological innovations that yield both bottom-line and top-line returns. This has only escalated further during the last decade. The majority of businesses are now well aware of the importance of sustainability, which represents a key driver for innovation and a challenge in many aspects at the same time. The challenges include not only obvious aspects such as products, processes, technologies and business models but also more abstract dimensions like cognitive, psychological and organizational challenges (Sharma 2017).

Sustainability in the agri-sector is of vital importance since this large sector uses over 37% of the total land area of the world and 21% of the land area of the European Union. If forestry is included, the area used reaches 68% of the world and 46% of the European Union (FAO 2019). Further, the agri-sector represents a significant component of the EU economy, accounting for about 7% of its total exports. The agri-sector also accounts for 4.3% of the total labour market in the European Union (Eurostat 2018).

Many businesses in the agri-sector are small companies with primary production as the dominating aspect of their business. These companies need new ideas and approaches to become more profitable at the same time as they are exposed to both internal and external pressure to become more sustainable. Internal pressure to build a business on sustainable values may arise from the board, management, shareholders, family and employees. External pressure can be derived from competitors, customers, special interest groups and governmental regulation-legislation (Tell et al. 2016).

There are also actors and forces working in the opposite direction, contributing to an increasing "unsustainability" of the agri-food sector. Bernard et al. (2014) have identified actors whose impact can lead to decision-making in the agri-company which leads to unsustainability: (i) loss-making investors and credit providers who abandon farms due to low economic returns; (ii) angry neighbours and environmental activists engaging in silent or active conflict, because they are negatively affected by farming activities; (iii) dissatisfied customers at the endpoint of value chains who do not trust the quality of products or disapprove of production conditions; and (iv) overacting regulators who over-regulate farm activities. Even though each actor perceives their actions as sustainable, they can influence agri-companies' management path towards unsustainability. Further, Fritz and Matopoulos (2008) have identified forces that can lead to unsustainability, such as (i) globalization of the agri-food industry resulting in increased imports and exports; (ii) consumer changes in consumption, resulting in a larger demand of food products, often out of season, that are transported long distances; (iii) the concentration of the sector, which has resulted in an ever-increased power imbalance in favour of retailers; and, finally (iv) major changes in delivery patterns with most goods now routed through supermarket regional distribution centres using larger heavy goods vehicles.

Despite the pressure to raise efficiency and lower costs from strong actors in the food value chain, many agri-food companies strive to conduct a sustainable business from social and ecological perspectives as well as from an economic one. Recent research (Cagliano et al. 2016) has identified three main integrated challenges for sustainability in the agri-food sector: first, the interdependency between food production and environmental, human and physical resources; second, the important role – sustainability and health aspects – of food for humans; and, third, the special characteristics of the food supply chain, with companies of different size and different sustainability focus.

The mindset and awareness of the owners and/or the managers of agri-companies can be an important factor for the development of sustainability-oriented innovation (Cagliano et al. 2016). Walker has identified this as a values-based driver (Walker 2012, 2014). Further, Barth et al. (2017) have found that many agri-food producers have a strong "value intention" to conduct their agri-business in a sustainable way. Explanations for this could be that many agri-companies are family businesses, rooted in their communities and strongly connected to the land of the ancestors. The owners/managers have experienced the effect of their actions on their land and production. They have accepted a responsibility for coming generations (Ulvenblad et al. 2016; Barth et al. 2017).

5.2.2 Sustainability-Oriented Innovation (SOI)

During the recent decades, the use of concepts connected to sustainable innovation, such as green innovation, environmental innovation and ecological innovation, have grown (Schiederig et al. 2012). In recent years, circular economy (Korhonen et al. 2018) and circular innovation (Guzzo et al. 2019) have also emerged as concepts. Bigliardi and Bertolini (2012, p 400) offer three explanations for this growing interest: "it may confer legitimacy, enhance competitiveness, and highlight ecological responsibility in an environment of both regulatory and consumer sensitization". However, the connection between innovation and sustainability has still to be developed further (Neutzling et al. 2018). Adams et al. (2016) have, after conducting a structured literature review of both academic and grey literature, identified four shortcomings with previous work. There is uncertainty regarding what sustainability actually means and how it can be achieved, because of the large variety of different conceptualizations. Previous work also tends to treat sustainability dichotomously (sustainable/not sustainable), rather than as a dynamic, unfolding process that is achieved over time. Further, previous work often overlooks the social dimension. Finally, many reviews of environmental management and sustainability exclude contemporary grey literature.

Based on their literature review, Adams et al. (2016) developed a framework of sustainability-oriented innovation. Their perspective is that "sustainability-oriented innovation involves making intentional changes to an organization's philosophy and values, as well as to its products, processes or practices to serve the specific purpose of creating and realizing social and environmental value in addition to economic returns" *(p. 181)*. The framework starts as regulatory compliance with incremental change at the firm level and culminates with radical change at the large-scale systems level. The researchers claim that moving through the framework requires a step change in philosophy, values and behaviour, which will be reflected in the innovation activity of the company.

The framework is divided into three dimensions (from diminishingly unsustainable to increasingly sustainable). The first dimension, technical/people, is about a movement in the literature from a focus on technology, i.e. a "set of tools", to a recent focus on people-centred innovation. The second dimension, stand-alone/integrated, is internal and describes how "innovation for sustainable manufacturing has moved from end-of-pipe, stand-alone solutions to modes of practice that require sustainability to be more deeply embedded in the culture of the firm" (p. 183). The third dimension, insular/systemic, reflects the firm's view of itself in relation to a wider socio-economic system beyond the firm's immediate boundaries and stakeholders.

These three dimensions represent three sustainability-oriented approaches on the journey to a sustainable business, operational optimization, organizational transformation and system building that a firm can have when it comes to innovation objective, outcome and relationship, defining the sustainability of the business (Table 5.1).

Table 5.1 A simplified model of SOI

Approach	Operational optimization: doing more with less	Organizational transformation: doing good by doing new things	System building: doing good by doing new things with others
Innovation objective	Compliance, efficiency "Doing the same things better"	Novel products, services or business models "Doing good by doing new things"	Novel products, services or business models that are impossible to achieve alone "Doing good by doing new things with others"
Innovation outcome	Reduces harm	Creates shared value	Creates net positive impact
Innovation's relationship to the firm	Incremental improvements to business as usual	Fundamental shift in firm purpose	Extend beyond the firm to drive institutional change

Taken from Adams et al. (2016)

5.3 Business Models

Business models are descriptions of how companies create value through exploitation of business opportunities (Rosca et al. 2017). Business models can be regarded as structured management tools, which are considered especially relevant for success (Magretta 2002). The research regarding business models has been growing since the mid-1990s (Osterwalder and Pigneur 2012). In recent years, research has put a steadily increasing focus on business models (Wirtz et al. 2016).

Most of this research addresses the importance of business models for companies' competitiveness, renewal and growth (Chesbrough and Rosenbloom 2002; Johnson 2010; Lambert and Davidson 2013; Teece 2010). Companies can apply several business models simultaneously, regarding, e.g. different products or markets (e.g. Aspara et al. 2013; Casadesus-Masanell and Tarziján 2012).

Researchers have used a variety of business models' definitions and settings in their studies, from the single company to the entire value network (Johnson 2010, Zott et al. 2011, Osterwalder and Pigneur, 2012). Even though empirical research on business models in 1996–2010 has focused on media, information technology and biotechnology industries (Lambert and Davidson 2013), a newly conducted structured literature review shows that the number of articles published regarding business models in food production has grown during the last 5 years (Tell et al. 2016). However, Wirtz et al. (2016) state that the growing field of research for the business model is in a consolidation phase, which still contains research gaps and thus offers possibilities for future research. The researchers suggest that research about business models' forms and components should be empirically validated, since a certain heterogeneity regarding this research area's focus has been determined.

Several researchers have discussed the central building elements of a business model: (i) value proposition, (ii) value creation and delivery and (iii) value capture (Bocken et al. 2014; Richardsson 2008; Osterwalder and Pigneur 2005). The value proposition is typically concerned with the product and/or service offering to generate economic return (Boons and Lüdeke-Freund 2013). Value creation and delivery is at the centre of any business model, and companies create and deliver value by seizing new business opportunities, new markets and new revenue streams (Beltramello et al. 2013; Teece 2010). Value capture is about considering how to manage cost structure and create revenue streams from the provision of good, services or information to users and customers (Teece 2010); see Table 5.2.

5.3.1 Sustainable Business Models

This interest in social and environmental sustainability is not new. Thirty years ago the Brundtland Report called for sustainable development that meets "the needs of the present without compromising the ability of future generations to meet their needs" (WCED, p. 43). Many researchers argue that more leadership is still needed around the issue of social and environmental sustainability (e.g. Kurucz et al., 2017). Researchers have also stated that a narrow focus on profitability without more attention paid to social and environmental sustainability can even limit a company's achievement of its economic goals (Kiron et al. 2013; Schaltegger et al. 2016).

Regarding the development of the sustainability-oriented innovations field, an increasing number of scholars frame it as a business model challenge (Rohrbeck et al. 2013). Several researchers have stated that the business model concept is a productive way to study the creation and use of sustainable innovation, both in practice and in theory (Boons et al. 2013). Further, researchers have called for more studies of business models oriented towards sustainable development (Boons and Lüdeke-Freund 2013; Boons et al. 2013; Breuer et al. 2016; Stubbs and Cocklin 2008; Upward and Jones 2016). They have proposed alternatives to the traditional business model with its focus on maximizing growth and revenues and on minimizing costs. One alternative model is sustainable business models based on the network approach or the value-net approach (Breuer et al. 2016; Boons and Lüdeke-Freund 2013; Kähkönen 2012; Lawson et al. 2008).

Boons and Lüdeke-Freund (2013) state that research on sustainable innovation is lacking conceptual consensus, which is needed to further develop the field. Based on a review of previous research, the authors propose a generic business model concept with four key elements:

Table 5.2 Conceptual business model framework

Business model building element	Value proposition	Value creation and delivery	Value capture
Focus	Product/service Customer segments and relationships	Key activities, resources, channels, partners, technology	Cost structure Value streams

From Bocken et al. (2014), Richardsson (2008), Osterwalder and Pigneur (2005)

(i) Value proposition: what value is embedded in the product/service offered by the company.
(ii) Supply chain: how upstream relationships with suppliers are structured and managed.
(iii) Customer interface: how downstream relationships with customers are structured and managed.
(iv) Financial model: costs and benefits from (i–iii) and their distribution across business model stakeholders.

These four business model elements, when combined with a perspective on social and environmental sustainability, describe a sustainable business model (Boons and Lüdeke-Freund 2013). Organizations committed to such sustainability integrate their social, environmental and economic activities in order to create value for their customers and for society. The sustainable business model analyses not only how organizations produce and deliver goods and services but, at the same time, how they contribute to the improvement of society – environmentally and socially. A company, cooperation or other organization with a sustainable business model is often part of, to a greater or lesser extent, a community or region that highly values the sustainable society and the sustainable environment.

Because of this increased focus on social and environmental sustainability, many companies worldwide have taken a greater interest in sequential business model innovation in which they refine an existing business model or launch a new one. The business model canvas framework (Osterwalder and Pigneur, 2012) has been developed for companies to envision and implement sustainable business models in practice. One of the new tools promoted for this work is the strongly sustainable business model canvas (Jones and Upward 2014; Upward and Jones 2016). In practice, a stewardship style of leadership is required for use of sustainable business models in which leaders understand their role as temporary custodians of power. Such leaders are committed to achieving value for all organizational stakeholders, including society (Bocken et al. 2014; Harvey 2001).

When focusing on sustainable business models, Barth et al. (2017) have proposed that a fourth building element should be added to the previously defined building elements of business models, (i) value proposition, (ii) value creation and delivery and (iii) value capture (Bocken et al. 2014; Richardsson 2008; Osterwalder and Pigneur 2005), namely, (iv) value intention. Many agri-companies are owner-managed family businesses. The owners regard themselves as stewards or custodians of the company, the property and the environment, with a responsibility for living and non-living things (Ulvenblad et al. 2016, Barth et al. 2017). Research regarding sustainability-oriented innovation also stresses the importance of intentional changes to the philosophy and values of the organization (Adams et al. 2016). Including value intention of the owner-manager in the conceptual framework could present important insights of potential trade-offs and barriers when addressing growth ambitions based on social, environmental and economic aspects (Table 5.3).

Table 5.3 A conceptual sustainable business model framework including the value intention

Business model building elements	Value intention	Value proposition	Value creation and delivery	Value capture
Focus	Mindset of owner/ manager	Product/service Customer segments and relationships	Key activities, resources, channels, partners, technology	Cost structure Value streams

Developed by Barth et al. (2017), based on a framework developed by Bocken et al. (2014), Richardsson (2008), Osterwalder and Pigneur (2005)

5.3.2 *Eight Sustainable Business Model Archetypes*

The research framework, which was used to address the research questions stated above, namely, (i) "How do Swedish agri-companies apply sustainable innovation practices?", (ii) "Which sustainable business models do Swedish agri-companies use" and (iii) "How can these innovation practices and sustainable business models be understood?", is also based on Bocken et al.'s (2014) eight SBM archetypes. In turn, they build their archetypes on the nine business model "building blocks" of the business model canvas (Osterwalder and Pigneur (2012), Boons and Lüdeke-Freund (2013) reference. Among the blocks most relevant to our study are value propositions, key activities, key partnerships and revenue streams. Building on this key tool for the analysis of business models, Bocken et al. (2014) add sustainable social and environmental activities. They (p. 44) define SBMs as follows:

> Innovations that create significant positive and/or significantly reduced negative impacts for the environment and/or society, through changes in the way the organisation and its value-network create, deliver value and capture value (i.e. create economic value) or change their value propositions.

Table 5.4 presents Bocken et al.'s (2014) eight SBM archetypes. These archetypes can be used in order to identify patterns and attributes that facilitate the categorization of the business model innovations for social and environmental sustainability. The archetypes can constitute a base for the development of a common language for the development of sustainable business models in research and practice.

5.3.3 *A Combined SOI and SBM Archetypes Framework*

In this chapter, the framework developed by Adams et al. (2016) focusing on sustainability-oriented innovations (SOI) is integrated with the framework containing eight different SBM archetypes developed by Bocken et al. (2014) (see Table 5.5). The idea behind combining these two frameworks, which was developed by Ulvenblad et al. (2019), is to categorize the business model innovations and study the organizational development (the sustainability-oriented innovation practices and processes) taken by agri-companies in Sweden.

Table 5.4 Eight SBM archetypes

Sustainable business model archetypes	Description and operationalization
1. Maximize material and energy efficiency	Do more with fewer resources, generating less waste and emissions and fewer pollutants
2. Create value from waste	Eliminate "waste" by turning waste into useful and valuable input in other production activities, making better use of underutilized capacity
3.Substitute with renewables and natural processes	Reduce the environmental impact and increase business resilience by addressing resource constraints associated with renewable resources and man-made artificial production systems
4.Deliver functionality, rather than ownership	Provide services that satisfy the users' needs without having to own the physical products
5. Adopt a stewardship role	Pro-actively engage with all stakeholders to promote their long- term health and well-being
6. Encourage sufficiency	Identify solutions that will reduce consumption and production
7. Repurpose the business for society/ environment	Prioritize the delivery of social and environmental benefits rather than economic benefits (i.e. shareholder value) through close integration between the company and local communities and other stakeholder groups. Recognize that the traditional business model in which the customer is the primary beneficiary may shift
8. Develop scale-up solutions	Deliver sustainable solutions on a large scale to maximise benefits for society and the environment

Taken from Bocken et al. (2014)

Table 5.5 A combined SOI and SBM archetypes framework

Approach	Operational optimization: doing more with less	Organizational transformation: doing good by doing new things	System building: doing good by doing new things with others
Sustainable business model archetypes	1. Maximize material and energy efficiency	4. Deliver functionality, rather than ownership 5. Adopt a stewardship role 6. Encourage sufficiency	2. Create value from waste 3. Substitute with renewables and natural processes 7. Repurpose the business for society/ environment 8. Develop scale-up solutions

5.4 The Case of Sweden

Sweden is often categorized as a country leading sustainable agri-production in many areas such as environmental awareness, animal welfare, low use of antibiotics and access to high-quality natural resources (Bucht 2016). The Swedish Ministry of the Environment presented as early as 2003 a vision for sustainable development that strongly recommends all policy decisions take into account the longer-term economic, social and environmental implications (Swedish Ministry of the Environment 2003). This is a vision that applies to food producers in Sweden.

The OECD report regarding Innovation, Agricultural Productivity and Sustainability in Sweden (OECD 2018) identifies Sweden as one of the earliest OECD countries to raise awareness of environmental issues and develop environmental policies. The result has been that the negative environmental impact has decreased, although agricultural production in Sweden has remained stable. OECD (2018, p. 12) states that: "Swedish legislation, which reflects consumer and citizen preferences, sets norms and standards for food safety, environment and animal welfare that is well above EU requirements in many areas of agriculture and horticulture. Swedish consumers and citizens have a high level of confidence for the Swedish agricultural and food system". The majority of Swedish agri-companies regard social and environmental issues as part of their goals, besides economic revenue (Ulvenblad et al. 2019).

However, the agri-sector in Sweden is facing challenges as well. The Swedish agri-sector has undergone significant structural changes in the last 20 years. The surviving food producers have become larger through internal growth and/or mergers and acquisitions (Swedish Board of Agriculture 2018). Others have been forced into subcontractor roles with diminished managerial influence on production goals and activities. This power asymmetry, combined with low-quality business relationships, can lead to suboptimization of resources and a reduced capacity to identify and satisfy consumer needs (Benton and Maloni 2005; Schulze-Ehlers et al. 2014). Furthermore, some larger companies in higher positions in the food value chain may not share smaller companies' interest in social and environmentally sustainable innovation. Even when such sustainable innovation (sometimes in response to consumer pressure) improves a product's quality, these larger companies may be disinclined to adopt the innovations for use in their production activities, delivery systems and product portfolios. They hesitate primarily because of fear of greater logistics complexity and higher costs. In some instances, however, these structural changes in the agri-sector have resulted in more cost-effective production and distribution systems, although with survival and profit still greater concerns than social and environmental issues. Thus, many stakeholders (e.g. consumers, consumer rights organizations, the media and citizens) are asking for healthy food and more sustainable social and environmental innovation in the agri-sector.

This pressure from stakeholders combined with the situation with declining profits in spite of production efficiency and economies of scale has led many agri-companies to develop their business models towards sustainability and high-quality products. Further, many agri-companies try to advance in the agri-value chain and get closer to the final customer.

To summarize, it seems likely that Sweden, as one of pathfinders in the world regarding sustainability and innovation, can contribute in the strive towards an innovative and sustainable agri-sector. Considering the global needs and challenges, it is important to deepen the international cooperation regarding the development of sustainable business models in the agri-sector. Hence, by studying the Swedish context, we can identify barriers, challenges and possibilities that can be relevant in other countries as well.

5.5 Methodology

In this paper, I will present, discuss and analyse eight different agri-companies/agri-cooperatives connected to agriculture in Sweden. Since forestry is an integrated part of agri-companies of Sweden and often an important part from cash flow and solidity perspectives, I have also studied one organisation, a large cooperative, from the forestry sector. All the companies/cooperatives are presented in Table 5.6.

The empirical data consists of both primary and secondary data, which have been collected by a set of different methods. The primary data consists mainly of interviews with owners/managers or other representatives of the companies and visits on the company/farm. Initially, the respondents were asked to tell their story of the companies in their own words. A semi-structured interview guide with open-ended questions was also used. It covered subjects like company history, past and current business activities, customers, partnerships, networks, goals, culture, values, sustainability, innovation and business models.

The secondary data has been gathered through document studies (official economic records, printed material and Internet pages). This multi-method approach has been used in previous research on business models (Täuscher and Laudien 2018; Zott and Amit 2008). In the study presented in this chapter, multiple methods have been used for each case, but not all methods have been used for all cases. The collected data, including the interviews, have been analysed through categorizations and content analysis (Täuscher and Laudien 2018).

5.5.1 Högared Milk Farm

Högared's main business is to produce and sell milk to a large milk distributor higher up in the value chain. Currently, there are 190 milk cows at the farm. The company is owned by two brothers with their families. Besides the two owners, there are six employees at the company. The company has historically experienced few possibilities to develop and change the business model. However, during the last decade, the owners/managers have prioritized their competence building through several leadership and management courses and have now formulated a new vision for the company:

Table 5.6 Companies/cooperatives in the empirical study

Company	Main business	Other businesses	Organizational form	Employees Number	Turn over Euros	Operating profit Euros
Högared	Milk farm	Machinery leasing Farm shop (milk)	Limited company Family-owned	10	1. 5 million	0.15 million
Gäsene Dairy	DairyCheese	Farm shop (milk and cheese)	Cooperative Owned by 28 farmers (23 active milk producers)	35	18.0 million	0.9 million
Gudmund Farm	Farm shop Charcuterie Pig farming	Restaurant Educational courses	Limited company family-owned	12	1.6 million	0.24 million
Ästad Vineyard	Vineyard Hotel	Restaurant Adventure tracks	Limited company Family-owned	39	4.6 million	0.43 million
Wapnö Farm	Milk Cattle	Restaurant Hotel	Limited company Family-owned	85	14.6 million	1.6 million
The South Forest Owners (Södra)	Forestry	Bioenergy	Cooperative 51,000 forest owners	3,500	2.1 billion	0.19 billion
Green Farms	Cattle farming Charcuterie	Distribution and sale from associated cattle farms	Limited company Family-owned	7	2.0 million	−0.9 million
The Farmers (Lantmännen)	Cereals Agri-food	Sales of machinery and vehicles Bioenergy	Cooperative 35,000 farmers	9,850	4.0 billion	62.8 million

> Our vision is to be a well-functioning farm for animals and humans, a modern machine station with the customer in focus. As a company, we want to build a good reputation in our district.

They have diversified their business model and activities in several directions; they have started a mechanical workshop and a custom for hire service. In the workshop, farming companies and other companies can buy services as maintenance and repair of vehicles and farming machinery. In the custom for hire service, the customer can buy services such as harvesting, applying fertilizers and pesticides, etc. Recently, the company has diversified even more, starting to sell a small fraction of its milk production directly to end consumers in their farm shop and in some of the larger groceries in the neighbouring city.

The owners have developed the sustainability focus of the company step by step. They started their journey towards sustainability according to the archetype "maximize material and energy efficiency", which still remains as the dominating archetype. It also means that from a sustainability-oriented innovation perspective, they are focusing on operational optimization ("doing more with less").

The development of their sustainable business model is underway. A minor part of their business model fits the archetype "deliver functionality, rather than ownership". Further, the managers have continuously developed their stewardship role over the last years (which could be seen in the vision of the company). The company's sustainable business model is moving towards organizational transformation ("doing good by doing new things").

5.5.2 Gäsene Dairy

Gäsene Dairy is a dairy cooperative company owned by 28 small- and medium-sized milk farms (between 30 and 500 milk cows on each firm). The dairy's main business is production and selling of high-quality cheese.

The dairy was founded in 1930, when milk prices were low and it was difficult for the producers to get good enough prices. Farmers in one neighbourhood developed a new business model before the concept was even conceptualized. They joined together in a cooperative association and started their own dairy, which produced and sold milk, cheese and other dairy products. The vicinity was important for the founders of the dairy, and it is still important today. All the milk are produced on farms which are situated within 25 minutes travel time from the dairy. Most of the milk and cheese are sold in groceries in southern Sweden. A minor part is sold directly to end consumers at the dairy. The quality of their products, based on sustainability and closeness, has made their brand well-known. Last year, the dairy had over 100 bus loads with visitors, exceeding 23,000 visitors. The dairy has recently decided to use biofuel for heating its premises. The surplus heat will be used for heating of the neighbouring municipality senior housing.

Since the dairy products create larger value for end customers than the products sold by large international processing companies, the dairy company can sell their products at higher prices. Consequently, the farms owning the dairy have larger revenues than other farms which deliver to the large international processing companies.

The business model of the company as such has not changed much since the company was founded, but the development of society has renewed it. From the start, the business model was based on economic necessity but also on the founding farmers' stewardship perspective. A business that was regarded as out of date has now become both modern and sustainable. One of the owners says:

> We have been out of fashion for 70 years, but now we are modern and in the front again.

Since sustainability became an important societal concept during the last decade, the company focuses even more on sustainability. The investment decision regarding the biofuel heater, which will benefit both the company and the senior housing, is one indication of the sustainability focus. Further, the company has changed its communication with customers and emphasized values as quality, vicinity and sustainability.

The founding and succeeding farmers have over time acted based on a "stewardship perspective". Due to the societal change, the dairy matches the SBM archetype "repurpose the business for society/environment". It also matches "substitute with renewables and natural processes".

From a sustainability-oriented innovation perspective, their company has covered all three dimensions. Even though the dairy was founded in order to reach operational optimization (doing more with less), the stewardship perspective with focus on vicinity and sustainability in combination with societal change has led to organizational transformation and system building.

5.5.3 The Gudmund Farm

The Gudmund Farm is a farm charcuterie, which also conducts pig breeding and production. The company was founded in 1998 and produces high-quality sausages and other meat products. Fresh and processed meat products are sold in the farm shop, in other farms' shops and in the groceries in the city. The meat comes from the farm or from subcontractors, farms in the neighbourhood. The company has long-term spoken agreements with the subcontractors, based on trust and a handshake. The sausages are handcrafted by old methods and have no additives other than natural spices and herbs. The company has diversified its products and activities continuously over the years. The company develops new sausage varieties and other meat products. Some of the new products have been developed by the employees of the company. Recently, the owners have also started to provide courses in sausage craftsmanship, food waste minimization and nutrition. Even though one goal of courses is to generate some revenue, the main reason is to educate customers. The company has also started a restaurant at the farm, where they serve lunch and arrange conferences. In order to get closer to the end customers and to learn their needs and expectations, the company opened a shop in the city close to the farm charcuterie, in 2005.

An important part of the business model is to develop and nurture long-lasting cooperation with customers, other companies, subcontractors and neighbours. The owner is also explicit regarding sustainability:

> Our company and our farm will stay where it is. Of course, we have to take good care of our land, our employees, our neighbours and our animals. We also want to create win-win relationships with customers, sub-contractors and other companies.

The company matches three SBM archetypes: (i) encourage sufficiency, (ii) substitute with renewables and natural processes and (iii) adopt a stewardship role.

From the start, the company has been run by the owner with a clear and explicit stewardship perspective, where sustainability is a central theme. The owner stresses the importance of local and professional networks. The company was the first to apply this perspective when it started the business 20 years ago. Since then other companies have followed suit. From a sustainability-oriented innovation perspective, the company is conducting system building through the strategy of working in networks with a win-win focus.

5.5.4 Ästad Vineyard

Ästad Vineyard used to be a traditional farm, with 100 hectares of fields, meadows and pastures, producing milk and grain. The previous owner, and father of the current owners, changed to ecological milk production in the mid-1980s in order to raise the low profitability of the farm. That was the starting point for a continuous and ongoing sustainable business model development. The next step was to invite school classes with fifth graders to come and have an experience of the farming activities. It developed into a team-building concept, where companies could bring their employees to the farm and solve intellectual and practical problems together.

The current owners, three siblings, improved the old farm buildings and built some additional buildings, which they used to develop the business model with a spa integrated with the small river, a restaurant, a conference centre, a hotel and a winery/vineyard. Today, they have 15,000 vines, and the wine is sold to distributors and directly to end consumers at the restaurant. An important and explicit building block of their business model is to use and develop the resources of the farm, the small river, the buildings, the vegetables, the wine, etc. The owners claim that:

> By experience we know that every challenge we meet also leads to new opportunities.

The family was conducting a traditional farm business from its inception. A desire to increase profitability and catch opportunities led to diversification and continuous business development. The sustainability aspect was based on partly an identification of the farm and land as a key asset from a business perspective and partly on a stewardship perspective. The company matches three SBM archetypes: (i) deliver functionality, rather than ownership, (ii) substitute with renewables and natural processes and (iii) adopt a stewardship role.

From a sustainability-oriented innovation perspective, this firm is at the organizational transformation level (“doing good by doing new things”).

5.5.5 Wapnö Farm

Wapnö Gård is an estate with an old history. It has been one of largest farming estates in Sweden since the fourteenth century. The current owner’s family has owned Wapnö since 1741. Today, Wapnö is organized as a limited company and has about 85 employees.

Wapnö Farm used to be an ordinary, although large, farm producing milk. The owners and management regarded the farm as a producer at the onset of the agri-food value. The milk was delivered to a large organization, which now has become an actor on the international market. Over 20 years ago, the owner and the management decided to start developing a diversified sustainable business model. They choose to advance in the agri-food value chain and get closer to the end consumer. Wapnö focused on sustainable environment and the preferences of the end customers, e.g. taste and flavour experiences. Today, the management of Wapnö talks about "sustainability into the future". Wapnö has been working actively towards a sustainable environment for many years. The manager states that:

> the present generation should take care in using natural resources reasonably and being environmentally responsible so that we leave the environment as untouched as possible for future generations.

Wapnö is developing a circular economy with a diversified sustainable business model. Wapnö call themselves an open farm, which means that consumers can come to the farm to get a closer look at the animals, the barns and the dairy. Wapnö also has a restaurant, which uses ingredients from the farm.

The company has established a farm brewery, and the beer is brewed from the farm's water and grain. The cattle are moving freely and never given antibiotics. The cattle are not given soy products, but rather pressed canola. Wapnö is also producing biofuel from canola oil. The milk flows directly in a tube from the barn to the dairy 30 meters away. Wapnö also has a large greenhouse, where they grew vegetables for sale in the farm shop and for the farm restaurant. The greenhouse is heated with renewable energy generated on the farm.

Wapnö farm biogas, produced from cattle manure, contributes to renewable energy in the form of electricity, heat and cooling, which is needed year round in the food premises. Wapnö only uses manure from animals on the farm for biogas production and has cut the energy consumption with more than 90%. The biogas plant also provides fertilization, which improves the fertility and value of the farmland.

Through the development of a sustainable diversified business model, Wapnö has climbed the value chain, got closer to the end consumer and developed a very strong brand. Therefore, it is able to sell its products at a higher price, which reflects the value end customers put on the products.

As many other farms, Wapnö used to be a traditional, although large, farm business in the start of the agri-value chain. Over the last 20 years, Wapnö has developed a diversified sustainable business model, and the development is still ongoing. Today, the company matches several business model archetypes. Wapnö maximizes its resources, creates value from waste and substitutes with renewables and natural processes. Further, the company has repurposed the business for society/environment since the sustainability and circular economy are in focus. Finally, and not least, the stewardship role is very clear and articulated.

From a sustainability-oriented innovation perspective, Wapnö is a good example of a company which has developed sustainability as a process. The company has developed from operational optimization to organizational transformation: doing good by doing new things. It could also be argued that the company has developed to system building. Although the company is mainly working as an entity, the farm is large and has advanced to a form of system building, where one part of the farm is supporting – and getting support – from other parts of the farm.

5.5.6 Green Farms

In the beginning, Green Farms used to be run as a traditional farm by the current owner's father. When the son, the current owner, took over the company, he wanted to focus on sustainability and converted it to organic production in 1989. At first, sales did not go according to plan and the cash flow was below expectations. He could not afford to feed the cattle with expensive concentrate, so he had to feed the cows with cheaper roughage, grass in different forms. This meant that his cattle grew slower and were older than normal at slaughter. The owner expected that the meat would then be of low quality and hard to sell at a good price. However, he soon realized that the meat was of very high quality. Since it also was produced in a more sustainable way than before, it could be sold to a premium price to customer wanting high-quality meat.

Twelve years later, in 2001 he started Green Farms. The business model was to create a network of farms that raised and feed cattle in the same way as the first farm. A new farm can get into the network after a trial period, if they meet the sustainable production requirements of Green Farms. The farms have to focus on animal health, sustainability and meat quality. The network members deliver their meat to Green Farms, which sells, through the Internet, and distributes high-quality meat to the end customers, restaurants, public kitchens and individuals. Green farms cooperate with around 40 sustainable cattle farms in the southwest of Sweden.

When the present owner took over the farm from his father, he wanted to transform it into a sustainable farm. He took a stewardship role from the start and substituted the production processes to more sustainable processes. From a sustainability-oriented perspective, he had developed his company through all three phases. Today, the company has developed into system building, since the company has developed a scale-up solution and engaged other companies in the system.

5.5.7 The South Forest Owners (Södra)

The farmers who owned forest began to organize themselves as a forest owner association in the beginning of the twentieth century. The southern and middle part of Sweden was almost a deforested country at that time, due to bad forest management and short-sighted profit-maximizing forest companies. In the beginning, the forest association provided advisory services regarding forest management. However, one significant issue for the small forest farmers was that the market was dominated by few large forest companies. Hence, the small forest farmers could not get fair prices for their product or for their forests. As a response to this situation, their cooperative soon developed to coordinate distribution, sales and processing of timber and other forest-based products. Over the years many small forest cooperatives merged, and the large cooperative association Södra was formed in 1938.

Today, Södra has 51,000 small forest owners as members/owners. A large majority of these forest owners are very engaged in sustainable forestry. Many of them have been engaged even before sustainability was a concept in research and regulation. In fact, over time some of the forest owners have refused to manage their forests the way the authorities required, since the forest owners believed they could manage their forests in more sustainable ways. Södra states that:

> the overall mission of the owners is to secure the provision for the members' forest raw material and promote forestry profitability through advice and support, so that the members' forests can be managed responsibly and with sustainability and to contribute to a market-based return on the forest raw material.

The timber from the forest farms is refined in Södra's industries for sawn and planed timber products, interior wood products, biofuel and pulp for the market in the market. Södra runs one of Europe's largest sawmill operations and is one of the largest producers of softwood pulp. Södra also produces textile pulp of hardwood. Their three pulp mills have almost fossil-free production and generate a large energy surplus. This bio-based energy is sold, among other things, as green electricity and district heating. Södra also owns manufacturing company, which produces one-family houses. Södra claims that they are focusing in innovation in order to develop new products, based on the renewable wood raw material.

Södra is emphasizing sustainable forestry and a sustainable forest value chain. This means, among other things, that efforts are taken to ensure that members' forestry is conducted using methods that ensure the production capacity of forest land and forests as well as conservation of ecosystem services and biodiversity. Södra mainly uses biofuels for the industrial production processes. The energy surplus is delivered in the form of electricity to the open market, district heating to places near pulp mills and sawmills and solid biofuels for heating plants. Through the industrial activities, the forest raw material contributes to the local community's conversion to a more sustainable energy use. Efficient use of the forest raw material from a material and energy perspective creates new conditions for sustainable products.

Even though Södra is based on small-scale forestry with strong local roots and local relations, it has ascended in the forest value chain and developed to a large international actor. A large majority of the small forest farms are run by owners who regard themselves as stewards. The stewardship perspective is also clearly communicated by their large cooperative Södra. Södra is applying several sustainable business model archetypes besides stewardship, maximize, create value from waste and develop scale-up solutions. Hence, Södra's sustainability-oriented process is covering all three steps, encompassing system building.

5.5.8 *The Farmers (Lantmännen)*

The Farmers in Sweden started to organize in the end of the nineteenth century, and they founded the national association in 1905. Today, it is an economic cooperative, owned by 27,000 Swedish farmers, and has grown to one of the largest actors in agriculture, food and energy in Northern Europe.

The Farmers' focus is to provide the members with seed, fertilizer, plant protection products and feed as well as to receive, store, refine and sell what farmers grow. Other important elements of the business are sales of forest, construction and agricultural machinery. The Farmers is the largest purchaser of grain in Sweden. They claim that they protect the earth's resources in a responsible manner and are included in the entire value chain from farm to table. Their business model strives to deliver sustainable products and new innovative solutions to customers while at the same time creating value for our owners and contributing to a viable agriculture.

The visions and goals of The Farmers are closely connected to innovation and sustainability.

In 2018, The Farmers (Lantmännen in Swedish) was named one of Sweden's most sustainable brands (Sustainable Brand Index 2018). The Farmers strive after viable agriculture, greener energy and a sustainable food chain. The Farmers state that,

> Together we take responsibility from land to table.... we lead the processing of arable land resources in an innovative and responsible manner for tomorrow's agriculture... we create a viable agriculture.

The Farmers shares several aspects with Södra. Many of the small farm owners have also sustainability priorities and have stewardship perspectives. The Farmers has also developed applying several sustainable business model archetypes besides stewardship, maximize, create value from waste and develop scale-up solutions. From a sustainability-oriented perspective, The Farmers are also involved in system building.

5.6 Analysis

The eight companies and cooperatives will be positioned in the combined framework of sustainability-oriented innovations (SOI) and different sustainable business model archetypes. See Table 5.7.

All of the companies analysed share a stewardship perspective on the business. All of them are also regarding themselves as entrepreneurs and business leaders, not only producers. Further, all of companies have over time developed and moved closer to customers in the agri-value chain. Another relevant aspect of their sustainable business models is that they have diversified their business models. Since these companies are depending on natural resources and often connected to one place, it is hard for them to develop scale-up solutions. However, one way to develop

Table 5.7 The combined framework of sustainability-oriented innovations (SOI) and different SBM archetypes

Sustainability-oriented innovation Approach	Operational optimization: doing more with less	Organizational transformation: doing good by doing new things	System building: doing good by doing new things with others
Sustainable business model archetypes	1. Maximize material and energy efficiency	4. Deliver functionality, rather than ownership 5. Adopt a stewardship role 6. Encourage sufficiency	2. Create value from waste 3. Substitute with renewables and natural processes 7. Repurpose the business for society/ environment 8. Develop scale-up solutions
Högared	Maximize	Stewardship Deliver functionality	
Ästad Vineyard		Functionality (experience) Stewardship	Repurpose
Gudmund Farm		Stewardship Deliver functionality	Substitute with natural processes
Wapnö Farm	Maximize	Stewardship	Create value from waste Substitute Repurpose
Green Farm		Stewardship	Substitute Develop scale-up solutions
Gäsene Dairy		Stewardship	Repurpose
The Farmers (Lantmännen)	Maximize	Stewardship	Create value from waste Develop scale-up solutions
The South Forest Owners (Södra)	Maximize	Stewardship	Create value from waste Develop scale-up solutions

scale-up solutions is to organize themselves into larger cooperatives, like Södra and Lantmännen.

All companies are closely connected to the real estate where they are situated. The companies, all of which are owner-managed, are family businesses in which the families expect to retain ownership for the foreseeable future. As family businesses strongly rooted in their communities, the owners are not concerned solely with growth and revenues. The owners think of themselves as stewards or custodians of the company, the property and the environment, with responsibility for living and non-living things. Cooperation in network structures or cooperatives is important for these companies. Trust, common values, other-orientation and win-win perspective are crucial concepts in the network structures.

5.7 Conclusions, Future Research Avenues and Practical Implications

The research presented in this chapter builds on, and adds to, previous research regarding sustainability-oriented innovation (Adams et al. 2016), business model archetypes (Bocken et al. 2014) and building blocks of business models (Barth et al. 2017).

The sustainability-oriented innovation framework regards sustainability as a continuous process that is developed and achieved over time (Adams et al. 2016). Their study has also shown how organizations can develop to become more sustainable. Further, they suggest that the development of sustainability-oriented innovation often starts with intentional changes to the values of the organization and as a response to regulation.

The analysis based on the cases of the study presented in this chapter deepens the knowledge regarding why agri-entrepreneurs develop the sustainability aspects of their business model. This study shows that many agri-entrepreneurs have a stewardship intention and want to develop and preserve their company, relationships and environment for the future and coming generations. Some agri-entrepreneurs applied the values of sustainability even before the concept was used in literature and discourse. The agri-entrepreneurs who strive for sustainability are often ahead of, or even in conflict with, legislation and policy when they develop their sustainable business models.

The aim of the eight sustainable business model archetypes developed by Bocken et al. (2014) is to "develop a common language that can be used to accelerate the development of sustainable business models in research and practice". In the study presented here, stewardship is a frequent and important sustainable business model archetype. The stewardship archetype seeks to "maximize the positive societal and environmental impacts of the firm on society by ensuring long-term health and well-being of stakeholders (including society and the environment)". According to researchers behind such archetype, it can preferably be used in combination with other archetypes. Based on the analysis in this study, the stewardship role is of paramount importance. However, it could be argued that stewardship should not be regarded as a business model archetype. Rather, it is an explanation or an incentive for developing sustainable business models.

The frequent use of adopting a stewardship role can be explained by the unique characteristics of the agri-sector. As Cagliano et al. (2016) have shown, there is a clear interdependency between agriculture and environmental, human and physical resources. Walker 2012 and 2014 have also pointed on the awareness of the entrepreneur as a value-based driver for sustainable business models. Ulvenblad et al. (2016) and Barth et al. (2017) have elaborated on this relationship. The owners/managers regard themselves as stewards or custodians of the company, the property and the environment, with a responsibility for individuals, animals and growing things. The company is often based on a farm, which has been owned by the ancestors before. The company is depending on the resources of the land, and it is going to stay where it is. Relations to neighbours and other companies are also important and have to be managed and maintained. The stewardship perspective is important when developing sustainable business models in the agri-sector.

Barth et al. (2017) have suggested that when studying the development of sustainable business models, the building block "value intention" should be added to previously developed building elements of the conceptual business model framework: (i) value proposition, (ii) value creation and delivery and (iii) value capture (Bocken et al. 2014; Richardsson 2008; Osterwalder and Pigneur 2005). Sustainability-oriented research also underlines the philosophy and values of the organization (Adams et al. 2016). Based on the cases presented in this chapter, the value intention is an important base for a sustainable business model in the agri-sector. Hence, it seems relevant to include the value intention element into the conceptual business model framework as well. Including the value intention of the owner-manager in the future theory building could present important insights of potential trade-offs and barriers when addressing growth ambitions based on social, environmental and economic aspects.

Future research might also examine how agri-companies innovate their sustainable business models when they introduce new products and engage in new business activities. It would also be relevant to further deepen the understanding of the connection with the special challenges in the agri-sector (Cagliano et al. 2016), the importance of the value intention (Barth et al. 2017) and value-based drivers for sustainable innovation (Walker 2012, 2014). The case study approach is well-suited for such studies.

Another relevant question to further investigate is a comparative analysis of the sustainable business model concept among industries. "Literature indicates that a wide range of traditional SMEs are still mostly focused on harvesting low hanging fruits by engaging primarily in incremental innovation" (Klewitz and Hansen 2014). The results from the agri-sector in Sweden show that there are companies in the agri-sector that optimize their operation with doing more with less, but the majority states that they focus on organizational transformation or even system building. Further, many of the owners/managers of the agri-companies adopt a stewardship perspective. Hence, many companies in other industries, and not only SMEs, can gain inspiration, insights and experiences and learn from the agri-sector.

5.8 Practical Implications

If agri-entrepreneurs and society focus on added value higher up in the value chain rather than only on efficiency at the inception of the value chain, it will be natural to emphasize the importance of agri-businesses for a sustainable society. Instead of a focus on the negative climate and environmental impact of production, focus can be on creating added value from climate and environmental perspectives.

A growing awareness and understanding of these values increases the opportunities for agri-companies to obtain higher prices for their products and/or services. Agri-companies can then focus more on how sustainable innovations can be developed. The climate and environmental perspectives will then become an opportunity and not a limitation for agri-companies and society.

Extension services focused on the agri-sector should reinforce the concept of agri-entrepreneurs as entrepreneurs instead of just producers. The efforts should focus on new, sustainable business models that encompass more links in the value chain than primary production. This means that education and counselling to businesses should focus on leadership, business development and innovation. The dissemination of knowledge should be conducted through coaching, in order to strengthen the competence and ability of the agri-entrepreneur.

5.9 Concluding Remarks

It is important for sustainable development to further emphasize the importance of innovations in the agri-sector, not only for the agri-companies themselves but also as a response to many of the challenges society faces. While social development has benefited major cities and urban centres, rural areas and agri-companies have substantial opportunities and resources to contribute to the solutions to many of our major social challenges today. These social challenges apply to several major and comprehensive issues such as:

- Climate and environment
- Integration, diversity and gender equality
- Labour and employment
- Access to housing
- The degree of self-sufficiency of society

Agri-companies and agri-entrepreneurs are often situated in rural areas. They can provide solutions to these challenges. Their owners are aware that their companies are connected to a village, a place. They are aware, sometimes intuitively, that the family, the farm, the place and the company will exist in the future - even after they leave business. Therefore, they have every reason to take care of relationships, neighbours, companies, land and animals. Agri-entrepreneurs are aware of their responsibility for future generations and have opportunities to contribute to environmental and climate solutions, which are some of today's major issues. Their mindsets and value intentions ought to be spread in society in general. These agri-entrepreneurs are stewards in the best sense of the word.

References

Adams R, Jeanrenaud S, Bessant J, Denyer D, Overy P (2016) Sustainability-oriented innovation: a systematic review. Int J Manag Rev 18(2):180–205

Alston J (2010) The benefits from agricultural research and development, innovation, and productivity growth. OECD Food, Agriculture and Fisheries Papers, No. 31, OECD Publishing, Paris

Alston J (2018) Reflections on agricultural R&D, productivity, and the data constraint: unfinished business, unsettled issues. Am J Agric Econ 100(2):392–413

Aspara J, Lamberg J-A, Laukia A, Tikkanen H (2013) Corporate business model transformation and inter-organizational cognition: the case of Nokia. Long Range Plan 45(6):459–474

Barnett ML, Salomon RM (2012) Does it pay to be really good? Addressing the shape of the relationship between social and financial performance. Strateg Manag J 33:1304–1320

Barth H, Ulvenblad P-O, Ulvenblad P (2017) Towards a conceptual framework of sustainable business model innovation in the agri-food sector: a systematic literature review. Sustainability 9:1620

Beltramello A, Haie-Fayle L, Pilat D (2013) Why new business models matter for green growth. OECD, Paris

Benbrook CM, Butler G, Latif MA, Leifert C, Davis DR (2013) Organic production enhances milk nutritional quality by shifting fatty acid composition: a United States-wide, 18-month study. PLoS One 8(12):E82429. https://doi.org/10.1371/journal.pone.0082429

Benton WC, Maloni M (2005) The influence of power driven buyer/seller relationships on supply chain satisfaction. J Oper Manag 23(1):1–22

Bernard F, van Noordwijk M, Luedeling E, Villamor GB, Sileshi GW, Namirembe S (2014) Social actors and unsustainability of agriculture. Curr Opin Environ Sustain 6:155–161

Beuchelt TD, Zeller M (2012) The role of cooperative business models for the success of smallholder coffee certification in Nicaragua: a comparison of conventional, organic and organic-Fairtrade certified cooperatives. Renew Agric Food Syst 28(3):195–211

Bigliardi B, Bertolini M (2012) Green innovation management: theory and practice. Eur J Innov Manag 15(4)

Bocken N, Short S, Rana P, Evans S (2014) A literature and practice review to develop sustainable business model archetypes. J Clean Prod 65:42–56

Boons F, Lüdeke-Freund F (2013) Business models for sustainable innovation: state-of-the-art and steps towards a research agenda. J Clean Prod 45:9–19

Boons F, Montalvo C, Quist J, Wagner M (2013) Sustainable innovation, business models and economic performance: an overview. J Clean Prod 45:1–8

Breuer H, Fichter K, L*ü*deke-Freund F, Tiemann I (2016) Requirements for sustainability-oriented business model development. Paper presented at the 6th International Leuphana Conference on Entrepreneurship, 14–15th January, Lüneburg

Bucht S (2016) A National Food Strategy for Sweden – more jobs and sustainable growth throughout the country. Short version of Government bill 2016/17:104

Cagliano R, Worley CG, Caniato FF (2016) The challenge of sustainable innovation in agri-food supply chains. In: Caniato FF, Worley VG (eds) Cagliano R. Emerald Group Publishing Limited, Organizing supply chain processes for sustainable innovation in the agri-food industry, pp 1–30

Casadesus-Masanell R, Tarziján J (2012) When one business model isn't enough. Harv Bus Rev 90(1/2)

Chesbrough H, Rosenbloom RS (2002) The role of the business model in capturing value from innovation: evidence from Xerox Corporation's Technology Spinoff Companies. Ind Corp Chang 11(3):529–555

Coad N, Pritchard P (2017) Leading sustainable innovation. A Greenleaf Publication book. Routledge, London

Dobermann A, Nelson R (2013) Opportunities and solutions for sustainable food production. Paper for the high-level panel of eminent persons on the Post-2015 Development Agenda, Sustainable Development Solutions Network, 15th January 2013

European Commission (2011) Communication from the commission to the European parliament, the council, the European economic and social committee and the committee of the regions, Horizon 2020 – The Framework Programme for Research and Innovation

European Commission (2016), Communication from the commission to the European parliament, the council, the European economic and social committee and the committee of the regions, next steps for a sustainable European future – European action for sustainability

Eurostat (2018) Agriculture, forestry and fishery statistical book. ISBN 978-92-79-94757-5, https://doi.org/10.2785/340432, Cat. No: KS-FK-18-001-EN-N

FAO (2011) The state of the world's land and water resources for food and agriculture (SOLAW) – managing systems at risk. Food and Agriculture Organization of the United Nations, Rome and Earthscan, London

FAO (2019) Food and agriculture data, FAOSTAT

Fritz M, Matopoulos A (2008) Sustainability in the agri-food industry: a literature review and overview of current trends. Conference: 8th International Conference on Chain Network Management in Agribusiness the Food Industry, May 2008, Wageningen

Fuglie K, Clancy M, Heisey P, MacDonald J (2017) Research, productivity and output growth. US Agric J Agric Appl Econ 49(4):514–554

Griggs D, Stafford-Smith M, Gaffney O, Rockström J, Öhman MC, Shyamsundar P, Noble I (2013) Policy: sustainable development goals for people and planet. Nature 495(7441):305–307

Guzzo G, Hofmann Trevisan A, Echeveste M, Hornos Costa JM (2019) Circular innovation framework: verifying conceptual to practical decisions in sustainability-oriented product-service system cases. Sustainability 11(12):3248

Harvey M (2001) The hidden force: a critique of normative approaches to business leadership. SAM Adv Manag J 66(4):36

IPCC (2019) Summary for policymakers. In: Special Report on Climate Change and Land. An IPCC special report on climate change, desertification, land degradation, sustainable land management, food security, and greenhouse gas fluxes in terrestrial ecosystems. IPCC, Geneva, Switzerland

Johnson MW (2010) Seizing the white space: business model innovation through growth and renewal. Harvard Business School Publishing, Brighton, MA, USA

Jones P, Upward A (2014) Caring for the future: the systemic design of flourishing enterprises. In The third symposium of relating systems thinking and design (RSD3), Oslo, pp 1–8

Kähkönen A-K (2012) Value-net – a new business model for the food industry? Br Food J 114(5):681–701

Kiron D, Kruschwitz N, Haanaes K, Reeves M, Goh E (2013) The innovation bottom line. MIT Sloan Management Review Research Report Winter 2013. Cambridge

Klewitz, J, Hansen, E (2014) Sustainability-oriented innovation of SMEs: a systematic review. J Clean Prod 65:57–75

Korhonen J, Nuur C, Feldmann A, Birkie SE (2018) Circular economy as an essentially contested concept. J Clean Prod 175:544–552

Kurucz, E, Colbert, B, Lüdeke-Freund, F, Upward, A, Willard, W (2017) Relational leadership for strategic sustainability: practices and capabilities to advance the design and assessment of sustainable business models. J Clean Prod 140:189–204

Lambert S, Davidson R (2013) Applications of the business model in studies of Enterprise success, innovation and classification: an analysis of empirical research from 1996 to 2010. Eur Manag J 31(6):668–681

Lawson R, Guthrie J, Cameron A (2008) Creating value through cooperation - an investigation of farmers' markets in New Zealand. Br Food J 110(1):11–25

Lobell DB, Burke MB, Tebaldi C, Mastrandrea MD, Falcon WP, Naylor RL (2008) Prioritizing climate change adaptation needs for food security in 2030. Science 319(5863):607–610

Magretta J (2002) Why Business Models Matter. Harv Bus Rev 80(5):86–92

Neutzling DM, Land A, Seuring S, Nascimento LFM (2018) Linking sustainability-oriented innovation to supply chain relationship integration. J Clean Prod 172:3448–3458

Nidumolu R, Prahalad CK, Rangaswami MR (2009) Why sustainability is now the key driver of innovation. Harv Bus Rev 2009:57–64

OECD (2016) Agricultural Outlook 2016–2025. OECD Publishing, Paris

OECD (2018) Innovation, agricultural productivity and sustainability in Sweden, OECD Food and Agricultural Reviews. OECD Publishing, Paris

OECD (2019) Agricultural policy monitoring and evaluation 2019 (summary). OECD Publishing, Paris

Osterwalder A, Pigneur Y (2005) Clarifying business models: origins, present and future of the concept. Commun AIS 15:1–25

Osterwalder A, Pigneur Y (2012) Business model generation: a handbook for visionaries, game changers, and challengers. Wiley, Hoboken

Richardsson J (2008) The business model: an integrative framework for strategy execution. Strateg Chang 17(5–6):133–144
Rohrbeck R, Konnertz L, Knab S (2013) Collaborative business modelling for systemic and sustainability innovations. Int J Technol Manag 63:4–23
Rosca E, Arnold M, Bendul JC (2017) Business models for sustainable innovation – an empirical analysis of frugal products and service. J Clean Prod 162 suppl, S133–S145
Schaltegger S, Hansen E, Lüdeke-Freund F (2016) Business models for sustainability: origins, present research, and future avenues. Organ Environ 29:3–10
Schiederig T, Tietze F, Herstatt C (2012) Green innovation in technology and innovation management– an exploratory literature review. R D Manag 42(2):180–192
Schulze-Ehlers B, Steffen N, Busch G, Spiller A (2014) Supply chain orientation in SMEs as an attitudinal construct. Supply Chain Manag 19(4):395
Sharma S (2017) Competing for a sustainable world. Building capacity for sustainable innovation. A Greenleaf publication book. Routledge, London
Średnicka-Tober D, Barański M, Seal C, Sanderson R, Benbrook C, Steinshamn H et al (2016) Composition differences between organic and conventional meat: a systematic literature review and meta-analysis. Br J Nutr 115(6): 994–1011
Stubbs W, Cocklin C (2008) Teaching sustainability to business students: shifting mindsets. Int J Sustain High Educ 9(3):206–221
Sustainable Brand Index (2018) Official report Sweden
Swedish Board of Agriculture (2018) Agricultural statistics
Swedish Ministry of the Environment (2003) A Swedish strategy for sustainable development: economic, social and environmental
Täuscher K, Laudien SM (2018) Understanding platform business models: a mixed methods study of marketplaces. Eur Manag J 36(3):319–329
Teece DJ (2010) Business models, business strategy and innovation. Long Range Plan 43(2–3):143–462
Tell J, Hoveskog M, Ulvenblad P, Ulvenblad P-O, Barth H, Ståhl J (2016) Business model innovation in the Agri-food sector: a literature review. Br Food J 118(6):1462–1476
Thornton PK, Schuetz T, Förcha W, Cramer L, Abreu D, Vermeulen S, Campbell BM (2017) Responding to global change: a theory of change approach to making agricultural research for development outcome-based. Agric Syst 152(March):145–153
Ulvenblad P-O, Ulvenblad P, Tell J (2016) Green innovation in the food value chain—Will Goliath fix it—or do we need David? In: Proceedings of the 61st International Council for Small Business (ICSB) World Conference, New York, 15–18 June
Ulvenblad, P-O, Ulvenblad, P, Tell, J (2019) An overview of sustainable business models for innovation in Swedish agri-food production. J Integr Environ Sci 16 (1):1–22
United Nations (2015) Global sustainable development report. United Nations, USA
Upward A, Jones P (2016) An ontology for strongly sustainable business models defining an enterprise framework compatible with natural and social science. Organ Environ 29(1). https://doi.org/10.1177/1086026615592933
Walker S (2012) The narrow door to sustainability–from practically useful to spiritually useful artefacts. Int J Sustain Design 2(1):83–103
Walker S (2014) Designing sustainability: making radical changes in a material world. Routledge, London
WCED (1987) Our common future. United Nations Environment Programme, Nairobi, p 54
Wirtz BW, Pistoia A, Ullrich S, Gottel V (2016) Business models: origin, development and future research perspectives. Long Range Plan 49:36–54
World Bank (2019) Data set. agricultural land data source. Food and Agriculture Organization, Washington D.C., USA
Zott C, Amit R (2008) The fit between product market strategy and business model: implications for firm performance. Strateg Manag J 29:1–26
Zott C, Amit R, Massa L (2011) The business model: recent developments and future research. J Manag 37(4):1019–1042

6

Inclusive Agri-Business and Innovation

Niek van Dijk, Nick van der Velde, Janet Macharia, Kwame Ntim Pipim, and Hiwot Shimeles

6.1 A Brief Description of 2SCALE

The Netherlands-funded "Toward Sustainable Clusters in Agribusiness through Learning in Entrepreneurship" (2SCALE) is an incubator program that manages a portfolio of public-private partnerships (PPPs) for inclusive business in agri-food sectors and industries and ultimately to improve rural livelihoods and food and nutrition security. 2SCALE offers a range of support services to its business champions (small to medium enterprises [SMEs] and farmer groups) and partners in eight countries across sub-Saharan Africa, enabling them to produce, transform, and supply quality food products. These products go to local and regional markets, including to base of the pyramid (BoP) consumers.

Project support includes technology transfer, capacity strengthening, and market linkages, as well as other key elements such as improving access to finance – vital – but often sorely lacking in the smallholder sector.

Its implementing partners are (1) BoP Innovation Center (BoPInc) which supports companies and entrepreneurs to develop relevant products and services for and with the base of the pyramid (BoP). Its mission is to develop commercially and socially viable business models and activities which include the people in the BoP as consumers, producers, and entrepreneurs. (2) The Inclusive Green Growth Department (IGG) from the Ministry of Foreign Affairs of the Netherlands, which aims to ensure universal access to and wise use of natural resources. IGG is responsible for the Dutch foreign policy on the following themes: climate, water, food

We would like to sincerely thank Mahamane Toure and Maryse Ago da Silva who, along with authors, generously provided pictures to this chapter.

N. van Dijk · N. van der Velde (✉) · J. Macharia · K. N. Pipim · H. Shimeles
BoP, Inc., Utrecht, The Netherlands
e-mail: vandervelde@bopinc.org

security, energy, raw materials, and the polar regions. (3) The International Fertilizer Development Center (IFDC), a public international organization addressing critical issues such as food security, the alleviation of global hunger and poverty, environmental protection, and the promotion of economic development and self-sufficiency. (4) SNV (Netherlands Development Organisation) is dedicated to a society in which all people, regardless of race, class, or gender, enjoy the freedom to pursue their own sustainable development. We focus on increasing people's incomes and employment opportunities in productive sectors like agriculture as well as on improving access to basic services such as energy, water, sanitation, and hygiene.

6.2 BoP Marketing and Distribution in 2SCALE

In this section, we introduce the concept of base of the pyramid (BoP) and explain why, from both a developmental perspective and a business perspective, it makes sense to do business with and for BoP consumers. In ensuing sections, we explain why 2SCALE focuses on marketing and distribution to the BoP, how pilots are shaped within the program to introduce this way of working to business champions, and, in turn, how we continuously learn from testing this particular approach.

6.2.1 *The BoP as a Developmental Challenge and Business Opportunity*

The term "base of the pyramid," or "bottom of the pyramid," was first coined by CK Prahalad in his groundbreaking publication entitled *The Fortune at the Bottom of the Pyramid: Eradicating Poverty Through Profits* (Prahalad 2009). At that time, much was unknown about this market segment, but the interest generated by Prahalad's book ensured that during the years thereafter, the BoP became a topic of interest for many researchers and later for companies and nongovernmental organizations (NGOs).

One of the main reasons the book gained so much attention was that it was the first time a rationale had been built on why the BoP market was important. From a business perspective, it was interesting to gain a deeper understanding of the global economy's lowest income segment – a segment of around 4 billion consumers who live on less than US $1,500 per year.

One of the key findings of the book was that if and when companies and other stakeholders want to engage with the BoP, it requires a completely different approach in comparison to other market segments; even more so, it would "require radical innovations in technology and business models."[1]

To grasp a sense of how substantial the BoP market is from a financial perspective and to better understand the breakdown of market potential per sector, the International Finance Corporation (IFC) and World Resources Institute (WRI) shared the results of a key study on the BoP in 2007. The study revealed that the total consumer market size of the BoP, which remained largely untapped by companies at the moment, was valued at a staggering $5 trillion. The study placed the total value of the BoP market for food products in sub-Saharan Africa at $215 billion. This constitutes a huge new market potential for food-producing companies and farmer organizations targeting the African market.

At the same time, most BoP consumers are a part of the same group as the 821 million people who are undernourished globally.[2] Therefore, targeting low-income food consumers makes sense not only from a business perspective but also from a developmental perspective. Undernourishment should not be tackled purely by developmental interventions, such as food aid, but by using market-driven approaches that build upon viewing this vast group of 821 million as BoP consumers. These consumers, as Prahalad stated, are the "aspiring poor" – a group of consumers with very little to spend but highly aspirational in their purchasing decisions, making them also highly critical consumers.

6.2.2 *Marketing Toward the BoP*

As we described above, the BoP is quite a different consumer segment compared to others and therefore requires a different approach – a different type of marketing. Typically, in marketing, the 4Ps (product, price, place, promotion) approach is used. Instead, at 2SCALE we use the 4As (acceptability, affordability, awareness, availability) to cater to the different challenges and opportunities companies face in marketing their products to the BoP. The 4As approach also provides a better base for analyzing the unique challenges that BoP consumers themselves face when, for instance, making purchasing decisions. Altogether, the 4As approach provides an opportunity for a more in-depth and consumer-centric analysis of what it means to market products with and for the BoP. To have a successful marketing approach for the BoP, all four dimensions must be taken into consideration:

- *Acceptability:* Ensuring that the product is accepted by the end consumer. It requires an in-depth understanding of the consumer's needs, dominant behaviors, and customs.
- *Affordability:* Offering products at a price point that meets the purchasing power of the BoP. Interestingly, it has been proven that consumers at the BoP are willing to pay a slightly higher price for nutritious foods, if they understand the products' benefits.[3]

---

[1] 2018 FAO The State of Food Security and Nutrition in the World.

- *Awareness:* Making sure consumers know about the product and its benefits. In the process of creating awareness, it is key to identify any trusted channels.
- *Availability:* Making the product available to the BoP by building specific distribution channels, often requiring an innovative approach to reach remote and more informally organized BoP markets.

The 4As are better suited for the position in which BoP consumers find themselves. Oftentimes they are located in areas farther from typical markets or commercial centers than other market segments. Therefore, availability is important. As the BoP is also less connected to "mainstream" media than other consumer segments, companies need to take extra steps to promote their products to BoP consumers. In addition, BoP consumers have relatively lower purchasing power, so affordability is even more important to them than to other consumer segments.

In relative terms, food represents for BoP consumers a much larger share of their expenditure compared to other consumer segments – i.e., in Nigeria BoP households spend over 70% of their disposable income on food and beverages.[4] Lastly, companies will need to ensure that their products and services fit with the perceptions, traditions, and aspirations of BoP consumers, and these can often be quite different from other consumer segments. Business champions will have to make a greater effort, or take a different approach, in their marketing and communication to ensure that BoP consumers also accept the product.

Even though the BoP market might be more difficult to reach and requires dedicated effort, it also holds great potential. BoP marketing and distribution strategies provide business opportunities to companies as well as an opportunity to improve the food and nutrition security of an important population segment. BoP consumers are usually not perceived or targeted as potential consumers. Companies often do not perceive this market segment as an opportunity for generating profit due to their relatively low purchasing power. On top of the huge untapped economic potential, there is also a major development opportunity in providing nutritious food products to BoP consumers.

6.2.3 *BoP Marketing and Distribution Within the Context of 2SCALE*

A common driver across all 2SCALE activities is the market, providing either supply chains with thousands of smallholder farmers as end buyers or food product value chains targeting rural and urban food consumers. Value chain development with specific attention to BoP consumers also must be market driven to be successful. An important distinction is that the market opportunity offered by the BoP is not always (fully) recognized by value chain actors. Many of 2SCALE's partners have been serving high-end markets or business-to-business (B2B) markets in their countries or regions and therefore do not necessarily know how to reach BoP consumers. In addition, the business case can be quite different, requiring the development of specific or new product propositions for BoP consumers. While individual BoP consumers may have little purchasing power and therefore be more inclined to buy small quantities of product (sometimes referred to as "Kadogo Economy"),[5] the number of consumers and their market share are what drive economies of scale and profits for value chain actors (Boxes 6.1 and 6.2).

Box 6.1 BoP Marketing and Distribution – Theme: Women's Empowerment

The empowerment of women within agribusiness value chains is one of the main goals of 2SCALE. As discussed throughout this paper, developing BoP marketing and distribution activities can create both entrepreneurship as well as employment opportunities for women, beyond more "conventional" opportunities in agricultural value chains, such as factory workers and casual laborers. Good examples of this are the Likie Ladies, micro-entrepreneurs who have formed a network of last-mile distributors for business champion GUTS Agro in Ethiopia, and the Danaya women's processor cooperative, which built a strengthened business network out of the BoP marketing activities in Mali. In the future, 2SCALE plans to develop more activities that also empower female BoP consumers, primarily the female heads of households, through behavioral change campaigns and other activities.

Box 6.2 BoP Marketing and Distribution – Theme: Access to Finance

One of the challenging dynamics in BoP marketing pilots is access to finance or, more specifically, access to working capital. For instance, the micro-entrepreneurs who are involved in the sale and distribution of BoP food products often have limited access to working capital, making it difficult for them to buy stock. This can limit the demand for food products from business champions. Also, BoP consumers do not always have a steady income, sometimes limiting their ability to buy food, and micro-entrepreneurs often are not willing to sell food products on credit.

(continued)

(continued)

In the near future, 2SCALE will look into opportunities related to working capital needs of micro-retailers to keep stock and consumer finance opportunities, among others, by learning from interesting initiatives in the same field, such as the collaboration between Twiga Foods and IBM in Kenya.

Photo credit: USAID Ethiopia, Creative Commons Attribution-NonCommercial 2.0 Generic – creativecommons.org/licenses/by-nc/2.0/ – Image cropped and brightened.

In 2SCALE in general, we work together with business champions to realize their inclusive business ideas by setting up partnerships that involve different actors across the broader value chain. Such a business champion is either a small to medium enterprise (SME) or a farmer organization. Their business ideas can focus on involving more smallholder farmers, doing more business with micro-entrepreneurs, reaching more BoP consumers, or any other ways to make their operations and their broader value chain more inclusive. By developing a partnership that involves a broader range of actors (including financial institutions, input providers, and others), systemic change can be realized.[6]

Because the concept of BoP markets and the opportunities they hold are still relatively new for most 2SCALE business champions, the BoP consumer segment is always introduced in a pilot setting, within the broader context of a 2SCALE partnership. In this way, the business champion can be introduced to the potential of this market segment and experiment with approaches on how to reach the BoP. In this they are supported by 2SCALE to lower the risks that come with developing new products or marketing approaches to reach a new consumer segment.

6.2.4 *What Have We Achieved So Far*

In the early stages of 2SCALE, the focus was mostly on developing and implementing a market research methodology to gain market information and insights on BoP food markets (focusing on specific crops) in different 2SCALE countries. Though this research was quite detailed, it did not necessarily lead to direct, actionable insights for 2SCALE business champions. The approach was therefore changed, leading to a more action-oriented approach that eventually led to the implementation of 24 different BoP marketing and distribution pilots. These pilots were quite diverse in nature, ranging from small pilots with small-scale groups in groundnut processing to large pilots with established SMEs in packaged processed food products. In total, through different pilot activities (from market promotions to new product development), 2SCALE delivered the following results:

- Thirty-seven new product propositions for the BoP were introduced by 24 different business champions.
- In total, over a million of these new products were sold, often within a short pilot time span, substantially increasing the turnover of the business champions.
- Close to 250,000 BoP consumers were reached through different market activations and communication campaigns.
- Business champions directly created over 200 new jobs through BoP marketing and distribution activities (i.e., as sales agents, micro-distributors, and marketers).

6.3 How BoP Marketing Activities Are Developed Within 2SCALE

Thus far, BoP marketing activities have been implemented in 2SCALE in the form of pilots. The process of designing and implementing a BoP marketing pilot always takes place within the broader context of a 2SCALE partnership. For most 2SCALE partnerships, a general notion of the relevance of BoP markets is included when the partnership idea is developed and designed, and later it is described in more detail in the partnership description.

Subsequently, when the partnership agreement is signed and the partnership is ready for implementation, activities to implement the partnership are identified and developed into an annual action plan. It is in this process of action plan development that the BoP marketing activities also take shape in more detail.

The development of these BoP marketing activities, within the broader context of partnership agreements and action plan development, roughly follows five steps (Fig. 6.1) which are implemented together with the business champion. Even though every BoP marketing pilot is different because of differences in markets, geography, business partners, etc., the process of building a BoP marketing pilot follows a common path.

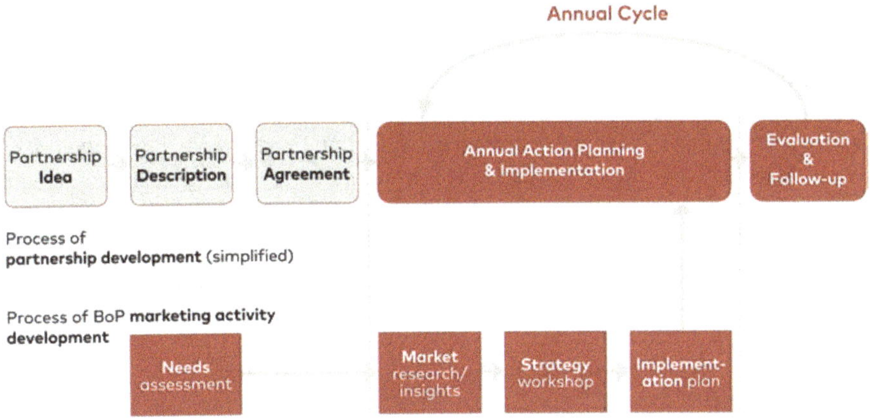

Fig. 6.1 Five-step process of designing and implementing a BoP marketing pilot

6.3.1 *Conducting Needs Assessments*

The first step in the process is to deeply engage with the business champion to get a joint and detailed understanding of his or her business and vision for targeting BoP consumers with nutritious food products. For the business champion to get a realistic understanding of what a BoP marketing pilot could look like in the context of 2SCALE, several cases of previous 2SCALE pilots are shared and described throughout this chapter. These practical examples often resonate well with the champion, since they include tangible learnings and also demonstrate the "business" results of targeting the BoP.

Another important part of the assessment is visiting the business champion's premises to gain insight about the current conditions under which the business champion is manufacturing products. There is not a formal assessment methodology that is used. Instead, specific attention is paid to the features that are relevant to designing the BoP marketing pilot. These features mainly relate to the champion's capacity in sourcing, production, marketing, distribution, and finance. For instance, it makes a large difference if an operational processing and packing facility already is in place. When such facilities are lacking, the 2SCALE teams know the implications in terms of required activities, timelines, and finance.

6.3.2 *Market Research and Insight Gathering*

The second step is to conduct market research and collect insights into consumer behavior in the particular market (and on the particular food product) that the business champion is active in.[7] These insights can also help provide an understanding of how a food product can improve the nutritional status of the targeted consumers, for instance, by making critical nutrients available or by making certain nutritional food products more affordable.[8] In addition, it is important that the market research and insights not become too generic and can provide actionable insights for the business champion to design the BoP marketing pilot. For example, one could question whether a small dairy cooperative that wants to develop affordable yoghurt requires extensive data on the national dairy market. Instead, the cooperative might be better off with a detailed understanding of the competitive yoghurt market in their specific location.

Overall, there are three important areas that are covered during this phase. First, there must be a general understanding of the market and the broader business environment. This should not be too generic but rather limited in its scope. Second, the competitive landscape will need to be understood in detail (e.g., which other companies provide similar products, what are their characteristics and pricing). Lastly, it is essential to have a detailed understanding of consumer characteristics, consumption, and purchasing behavior.

The goal of 2SCALE is to make the research phase as actionable as possible. Therefore, research is done in close collaboration with the business champion. The results of this phase are also actively discussed during the strategy workshop (see below), for instance, by purchasing and bringing along any competitive products to the workshop. The research and insights phase helps champions develop their BoP marketing strategy and design activities.

6.3.3 Strategy Workshop

After the research is finalized, all insights are brought together in a strategy workshop in which the key members of the business champion's management team are present. The most important aspect of this workshop is that it is co-creative, meaning that all content is being developed by the business champion's team together with 2SCALE.

The general approach of the workshop is to start mapping out the business champion's organization. To do so, 2SCALE uses the Business Model Canvas (BMC). This is a convenient model that has proven to be well understood by various business champions and provides actionable building blocks for a strategy. The most important building blocks of the BMC are the customer persona, value proposition, and marketing and distribution. For more information on these key components, please refer to Sect. 6.2. During the workshop, each of these building blocks is explained in an interactive way, ensuring that participants understand the content and relevance. After this, workshop participants are asked to build specific strategies for each building block in teams.

---

[2]An example is the soya goussi in Benin. This byproduct of soybean processing is full of protein and therefore a good alternative to poultry meat. Since the soya goussi is over 30% cheaper than poultry meat, it makes protein more affordable for BoP consumers.

Each workshop ends with a half-day discussion on planning the next key step: implementation. Once the general BoP marketing strategy is defined, activities are identified to implement the strategy, and a budget for these activities is established. This is input that will be integrated into the broader annual action plan that forms the main document for implementation of the overall partnership agreement.

6.3.4 Implementation

After intense discussions on market insights and strategies, time for action is due. During and after the strategy workshop, all activities and responsibilities have been formulated and formalized into the broader action plan and thereby form part of the broader partnership. It is also for this reason that the respective partnership facilitator[9] is an active participant in the abovementioned strategizing.

The BoP marketing expert and the partnership facilitator work closely together with the business champion in the implementation of activities. Sometimes there might be a need for specific external experts or marketing agencies. In this case, 2SCALE facilitates the process of bringing in such parties. These types of experts are often needed in technical product formulation (e.g., developing a recipe for nutritional porridge), marketing material development, or product certification.

There are three key milestones throughout the implementation stage. The first is to have the actual product ready. This sounds straightforward but, depending on the situation, it might be challenging to get the right ingredients, product formulation, machinery, or production setup in place. The second milestone is getting the packaging finalized. This includes having a clear value proposition, brand, communication strategy, and actual packaging design and finding the right party that has the right packaging material available. When these milestones are reached, the business champion has an actual physical product that is ready to be sold in the market and should suit BoP consumers' aspirations in terms of affordability and acceptability.

The third milestone is to launch a market activation campaign. This step has proven to be one of the most crucial components of implementation. Having an actual product ready for sales and distribution at the factory gate is just the start. There needs to be awareness among the consumers about the new product and it needs to be available in the market. To achieve this, the business champion and 2SCALE develop a market activation campaign that follows the ATEAR (attention, trust, experience, action, retention) framework that is highlighted in Sect. 6.2. In short, awareness is created by organizing several activities to promote the brand and product and engage local communities. Availability is created by working closely with local shopkeepers, social networks, and possibly sales agents to get the product to the consumer.

[3] A partnership facilitator is a 2SCALE staff member who is responsible for facilitating the overall partnership.

6.3.5 *Evaluate and Follow-Up*

The final stage of the process is to evaluate the pilot and to prepare the business champion for follow-up, which often consists of a strategy to either scale active marketing and distribution support by 2SCALE or exit the program and continue independently. This stage often starts with a meeting, during which the business champion reflects on the pilot, develops lessons learned, and shares ambitions toward the future. Based on that, together a follow-up strategy is developed. Looking back at the portfolio of pilots, there are four different outcomes of this stage:

- *Optimization:* The first outcome is when the business champion requests to continue to work together with 2SCALE to optimize the BoP marketing approach. This, for example, happened in the soybean partnership in Ethiopia with business champion GUTS Agro. During the evaluation, they realized that after successful product development and marketing, the next challenge was to safeguard the affordability of the product – a challenge that occurred mostly in their distribution model. Based on this workshop, GUTS Agro and 2SCALE designed a second phase of the pilot focusing on last-mile distribution.
- *Replication:* The second outcome is when the business champion sees potential to replicate the pilot activities in another geographical location but is not yet ready to do this alone. For example, with Shalem Investments in the sorghum partnership in Kenya, it was decided to replicate the pilot, which mostly focused on product launch and market activation. The pilot was originally implemented in Meru and was to be replicated in Thika.
- *Financing:* The third outcome is when the business champion sees opportunities for scale but is lacking the capital to finance growth of the model. In this scenario, 2SCALE supports the business champion in their business plan development and making relevant connections with the finance sector. For example, during our pineapple partnership with business champion Promo Fruits in Benin, 2SCALE supported Promo Fruits in developing a business plan to attract investments to scale its newly established micro-distribution model.
- *Exiting:* The fourth outcome is when the business champion feels sufficiently capacitated to continue independently. Eventually, all business champions will "exit" the program one way or another. However, that does not mean that all ties are broken. On the contrary, the business champion remains part of the 2SCALE network and often functions as a role model or mentor for other business champions.

6.4 Key Tools, Approaches, and Implementation Strategies

In this section, we present the tools and approaches that we use in the steps mentioned under Sect. 6.2 to develop implementation strategies for the pilots. Subsequently, we present the four generalized implementation strategies that we have seen materialize in the different pilots implemented over the past years.

6.4.1 Tools and Approaches

When building BoP marketing pilots together with the business champion during strategy workshops, several tools have proven to be of crucial value. These tools were adopted by 2SCALE from the Business Model Canvas and contextualized by the 2SCALE team. All of the tools are hands-on, practical, and co-creative so that they can be used in workshops and understood by low-literate participants. It is interesting to see that the same tools have worked for vegetable farmers in Ethiopia as well as for soybean processors in Benin.

6.4.2 Business Model Canvas

The Business Model Canvas (BMC) is a tool that maps out the various key components of an organization and shows how the various components work together (Osterwalder and Pigneur 2010). The primary value for business champions is that the BMC shows the entire layout of their business. Moreover, it shows how an additional activity, such as targeting the BoP, can have implications for other parts of the business. The goal is for business champions to understand their business, the implications of the pilot, and the required strategy to make a pilot successful.

The main components that are always used in workshops and that form the basis for pilots are as follows:

- *Persona:* A persona is a fictional character that represents a typical consumer that a business champion aims to target. Examples of features that are covered in the persona include consumer demographics, pains, and gains. The purpose of building a persona is to have an in-depth understanding of the persona's wants and needs, which helps the business champion in designing the value proposition.
- *Value proposition:* A value proposition is an elaborated description of the product or service that the business champion is selling to his or her consumer. More specifically, it shows how a product or service creates gains or reduces pains for the consumer. The overall goal is to create a product-market fit, ensuring that the value proposition is relevant for the target consumer.

6.4.3 Product and Pricing Strategy

As mentioned in the previous section, it is crucial for a business champion to have a clear and competitive product and pricing strategy in comparison to its competitors. To get to such a strategy, it is helpful to visit sales points and purchase competitive products. With the business champion, the competitive products are organized in terms of price and value. Finally, the champion has to determine where to position his or her (new) product. A good example comes from the sorghum partnership in Kenya with business champion Shalem Investments, which positioned its brand in the relatively lower segment in terms of pricing. However, Shalem used a premium plastic packaging material with a modern design and was able to do this at a relatively low price point. This has made the brand distinguishable from its competitors, which used more basic packaging with poorly designed product labels. Even though it sounds straightforward, this interactive exercise has proven its value in developing a clear product and pricing strategy.

6.4.4 ATEAR Marketing Model

This is one of the most important tools to build a BoP pilot and forms the foundation for a business champion to build an entire marketing campaign. In summary, the ATEAR model covers five crucial steps: attention, trust, experience, action, and retention (Fig. 6.2). During the workshop, the team ideates on each of the steps and builds a marketing campaign plan that follows all five steps in a structured way. The ultimate goal is that the business champion develops a coherent and comprehensive campaign instead of only focusing on one element of ATEAR.

- *Attention:* The first step is the most straightforward one – getting attention. If a new product is launched in the market, consumers need to be aware of it. To create this awareness, several activities can be organized. These can be common above- or below-the-line marketing activities, such as radio advertisements, market activations, or community messaging. In defining the activities, it is key to understand which channels reach the target customer and which are seen as trusted channels.

- *Trust:* Attention alone is not sufficient. Consumers need to trust a new brand. There are many different brands out there that seek attention, so how do you stand out and create trust? Trust takes time and can be built by consistently "showing up" as a company and delivering quality products. Having the right food safety certification or product endorsement by a public or trusted person from the community can be a valuable asset to building trust.

- *Experience:* The proof of the pudding is in the eating. If a consumer cannot experience, or more specifically taste, the product, they will not immediately be inclined to purchase a new product, because the "risk" of buying something that, for instance, does not taste good or have the right texture will be too high. That is why tasting sessions are often held during market activations, allowing prospective consumers to try small samples of the product.

- *Action:* When the consumer is finally convinced to buy the product, he or she sometimes needs a final push to actually make the purchase. It can help if a particular promotion action is available, for instance, buy three items and pay for two, in which a consumer is rewarded for making a larger (and therefore more economical) purchase.

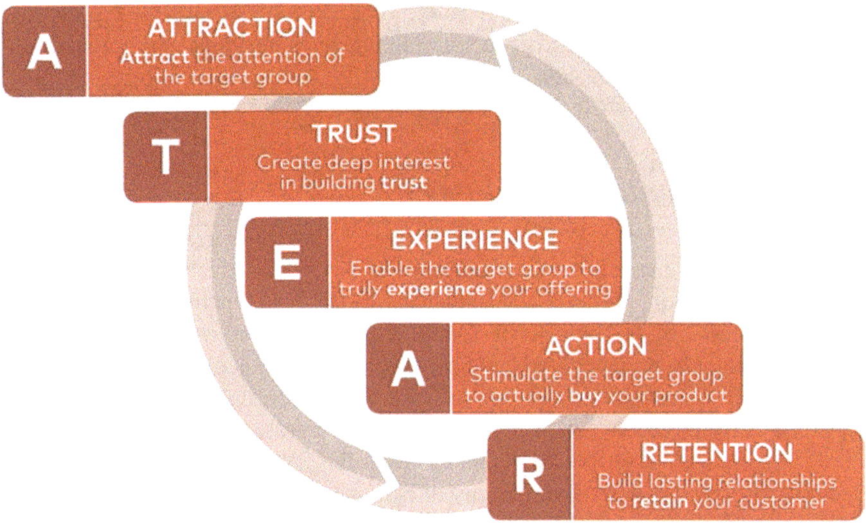

Fig. 6.2 ATEAR marketing model

- *Retention:* Getting new consumers to buy your product is essential, but in the world of fast-moving consumer goods (FMCGs), it is much more important to retain consumers and make sure they keep buying your product. Furthermore, it is much more expensive to create a new customer than to keep an existing one. Even though the short time span of BoP marketing pilots did not always allow for implementation of retention activities, loyalty activities were sometimes implemented, for instance, through cards that consumers can save to get a product for free.

All business champions have to compete in the market with other brands. Therefore, it is crucial that they come up with a brand and product that are distinct and take up a competitive position within the market. For this, 2SCALE developed a branding tool. In essence, the tool explains that a successful brand would need to:

- Resonate with the consumer
- Differentiate itself from competitors
- Express the internal values of the organization

For the first two criteria, the aforementioned persona and product and pricing strategies are used. For mapping the internal values, the first step is to define the core and aspirational values of the organization. These are summarized in the "brand (wo)man," having a single overview that captures the most important values of the organization. Secondly, a brand-archetype model is used, featuring eight different archetypes. An archetype is a typical representation of a certain set of personal traits and summarized in a set of "emotional values." After these steps are covered, the organization has a clear sense of its brand and communication strategy. This strategy provides clarity on the development of the organization's brand, packaging, and marketing materials. At the end of the session, the team develops conceptual directions for their packaging design.

6.4.5 BoP Distribution Modeling

Especially for business champions that have not yet been actively selling products in retail, distribution can be a key bottleneck to success. Distribution at the BoP level is a challenge for both local SMEs and cooperatives as well as for large multinationals. The main reason is that distribution and retail are not well organized at the BoP. BoP consumers are more difficult to reach and often remotely located from conventional markets or retail outlets, not just in rural areas but also in urban ones. This presents a serious challenge to reach the BoP consumer in an efficient and cost-effective manner. The majority of BoP consumers still make most of their purchases at open markets, local kiosks, or mom-and-pop stores – often owned by a single business owner. The BoP distribution model suggests three different approaches that business champions can explore within their specific market. These are as follows:

- Using existing channels and infrastructure (piggybacking)
- Creating hybrid partner-ships with nonprofit partners
- Setting up micro-franchised distribution models

The extent to which (a combination of) any of these three options can apply to a business champion strongly depends on two types of trade-offs (Fig. 6.3). One is the investment the business champion is able and willing to make in relation to the power or control the champion wants to have over the actual distribution activities.

The second trade-off relates to the speed at which the business champion wants to set up their distribution activities versus the level of competitiveness that the distribution model can bring to the actual product distributed.

The different characteristics of these three approaches are described in more detail under the following section on implementation strategies.

6.4.6 Implementation Strategies

Even though each pilot is unique and dependent on local market circumstances, we can distinguish four main implementation strategies for a BoP marketing pilot. Sometimes a pilot only consists of implementing one particular strategy; in other cases, several strategies are implemented in one pilot.

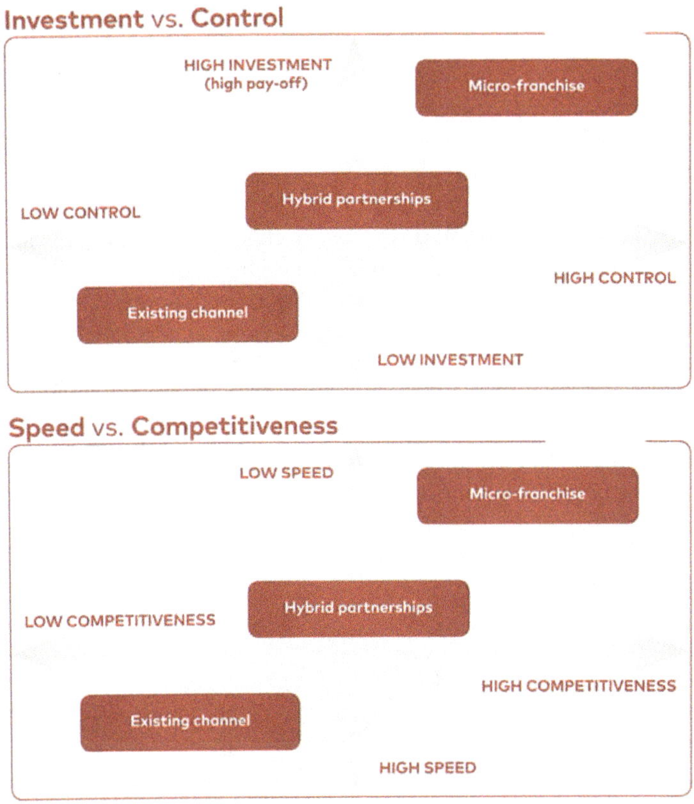

Fig. 6.3 Key trade-offs in the BoP-focused distribution models

6.4.6.1 Product Development

In the development of a pilot with a business champion, we often notice that after initial needs assessments and market research are conducted, the existing product portfolio of the business champions does not match the needs and wants of the local BoP consumer market. This means that 2SCALE will work with the business champion on developing a completely new product or will adapt the existing product in such a way that better suits the needs and wants of BoP consumers.

These processes can be quite time consuming and require a significant investment of resources from the business champion. But equally, they can provide a breakthrough for the company in entering the BoP market. In these cases, the market insights collected are used to shape the idea for a new product, and the product is subsequently further developed.

Sometimes this means that external expertise is needed, especially on the technical side. For example, a food technologist (i.e., through the Dutch "Programma

Uitzending Managers" [PUM] program[10]) was brought in to assist with one pilot. Also, at this point, we use some of the tools and approaches mentioned above to ensure that the product is appealing and aspirational to BoP consumers. Quite often, several iterations of the product are done, based on tasting sessions that are held with test panels of BoP consumers to better understand their appreciation of the product. Eventually, the product is launched or introduced on the (local) BoP market.

GUTS Agro in Ethiopia is a good example of a pilot where a completely new product was introduced (Box 6.3). In this example, the business champion was already producing a blend of corn and soybean for the World Food Program and other relief organizations. However, the company aspired to enter the BoP consumer market directly. To do this, it had to develop the product completely from scratch, based on the corn and soybean blend it was already producing in bulk. The market insights gathered in Ethiopia were used to develop a porridge formula consisting of corn and soybean and fortified with vitamins and minerals. This formula was developed in close collaboration with an expert from PUM. 2SCALE further supported the business champion to develop a new product name, tagline, packaging, and pricing strategy, leading to the introduction of the new "Supermom" product for the BoP market in Ethiopia.

Box 6.3 Business Champions – Snapshot: GUTS Agro, Ethiopia

GUTS Agro Industry is an ISO 222000-certified food-processing company in Ethiopia, producing a range of goods from cereal to table salt and baby food. 2SCALE has supported Engidu Legesse, CEO of GUTS Agro, to develop a micro-franchise distribution model called "Likie."

The Likie model is designed to effectively reach low-income consumers while maintaining the affordability of the product. The model uses female micro-franchisees to distribute a low-cost, high-protein, corn-soybean blend

(continued)

(continued)
known as Supermoms, also developed with 2SCALE assistance. The Likie distribution model is now operating in several areas in Ethiopia and is supported by marketing activities.

This five-day campaign featured three well-known musicians who led a promotional campaign in schools, marketplaces, and low-income neighborhoods. Every morning for a week, the musicians, accompanied by GUTS Agro staff and saleswomen, would visit a school. The infotainment included a quiz contest on nutrition, with GUTS Agro products as prizes.

Likie is not only about nutrition, but also about empowering women. "Likie Ladies" are provided with uniforms, branded bags, and a tricycle to transport their goods door to door. They receive training, business support, and stocks on credit.

For GUTS Agro, Likie is more than a distribution network. It is also a way to help improve health and nutrition among the poor while helping budding women entrepreneurs create new businesses with very low start-up costs.

In other cases, adjustments were made to existing products so that these products would better fit BoP consumer demands. In Benin, under the soybean partnership, 2SCALE worked with business champion Coopérative de Transformation, d'Approvisionnement et d'Ècoulement de Soja (CTAE) on the BoP marketing of a soybean-based product called soya goussi. CTAE was already producing and selling soya goussi, but the packaging was unappealing, and the product was prepared under unhygienic circumstances. Plus, the size of the product did not appeal to the consumer. Based on these insights, CTAE and 2SCALE came up with an improved way of producing the soya goussi, leading to a higher-quality product at the same cost, which was better appreciated by the consumer. It was also sold in larger-sized and higher-quality packaging, with a label that provided information on product characteristics, which was missing previously. Consumers responded positively to these changes, which contributed to a 15–20% increase in sales, compared to the original product.

6.4.6.2 Branding

Even though branding could be seen as part of a product development strategy, we describe it here separately, because the importance of good branding, particularly to the BoP, is often largely underestimated. Companies tend to refrain from investments in branding and use very basic packaging, promotion, and other branding materials – or do not have the capacity in-house to develop such branding materials. At the same time, the product itself not only needs to be of good quality or well priced, but it also needs to look good, and the broader brand needs to appeal to BoP consumers to help express the company's values. A successful brand meets three key criteria:

- Appeals to the customer
- Is distinct from the competition
- Represents the values of the organization

The relevant tools for this strategy are the persona, value proposition, and branding. As an example, 2SCALE supported Shalem Investments in Kenya in product and brand strategy, developing a nutritious porridge targeted at the BoP (Box 6.4). The first step was to develop a product formula that met the nutritional needs and taste palate of the Kenyan BoP consumer. Once the product formula was finalized, the key challenge was to develop a brand strategy and packaging design. To ensure the brand appealed to the customer, Shalem conducted consumer research. The key conclusion from this work was that customers were highly aspirational and concerned by the quality of the product. To meet these preferences, Shalem developed a modern and colorful brand in quality, printed pouches – ensuring that the product had a high sense of perceived quality. This strategy directly allowed Shalem to differentiate itself from its competitors, who were mostly still using printed paper bags with limited colors. Finally, Shalem had a strong vision for its organization that was incorporated in the new packaging design: embracing an energetic and empowering view on providing quality nutrition.

Box 6.4 Business Champions – Snapshot: Shalem, Kenya

Founder of Shalem Investments, Ruth Kinoti, is one of 2SCALE's Kenyan business champions. Prior to joining the 2SCALE program, Shalem focused on the aggregation and sales of farmers' produce, such as maize, beans, sorghum, and soybeans. During the 2SCALE program, Shalem transformed into processing and marketing goods under the brand name Asili Plus in the Kenyan retail sector.

Over the course of 3 years, Shalem has passed all five steps of the 2SCALE journey. At the core of this journey was the added value for sorghum targeted to consumers. We facilitated co-creation workshops to identify gaps in the market and consequently develop a branding and distribution strategy. The outcome was a re-branded porridge flour that was affordable for BoP consumers. The packaging was also designed and developed considering strict country regulations on plastic use.

(continued)

(continued)

To promote it and make it available, a marketing and distribution strategy was developed and implemented. In just a few short weeks after launching her new marketing strategy, using a small pool of promoters, Kinoti reached 800 women in the community. For market activation, a tent was built at one of the largest open markets in Meru, the Gakoromone market. There were music events, talks on nutrition, and 10 brand ambassadors, young women and men, who introduced the product to vendors and customers in the market.

Over 2000 people were reached during those market days, which took place once a week for 3 weeks. The customers had the opportunity to taste the Asili Plus porridge flour and received promotional items, such as free T-shirts, after purchasing more than four packets of the porridge.

Shalem Investments plans to grow its business by replicating the marketing and distribution strategy to other parts of the country, which will triple its income. The women involved during the pilot will be part of the micro-franchising model, in which they will continue selling the Asili Plus fortified flour as an income-generating activity.

Another example of brand development comes from our groundnut partnership located in the Ivory Coast. The business champion, the Komborodougou Women's Cooperative, previously sold unbranded groundnut paste in simple plastic bags. By selling the groundnut paste in this very basic packaging, it was difficult for consumers to distinguish their product from other similar products. Eventually, together with 2SCALE, a new type of packaging was selected – small plastic containers. Along with a newly developed brand, which was applied to the plastic containers, the brand concept characterized the region where the product was produced, making the branding also easily applicable to any other products the cooperative would develop in the future.

6.4.6.3 Market Activation

Earlier in this chapter, we introduced the ATEAR model as a critical building block for BoP market activation activities. These market activations are meant to bring (new) products to the attention of prospective consumers, and let them experience the product, ideally in the presence of a recognizable, trusted role model, to increase the likelihood that the prospective BoP consumers will trust the product. Market activations are often held at locations where BoP consumers typically come to buy their products (such as open markets during market days) or at places where peers meet or trusted people can be found. Examples of these are activations in women's groups (like Chamas in Kenya) and around churches or community halls in the presence of local religious or community leaders. Sometimes, activations are specifically targeted at a particular audience. In the case of the soy kebab pilot in North-East

Ghana, part of the broader soybean partnership, we identified the significant potential of selling nutritious snacks (soy kebabs) through local schools, instead of the sweets and cookies that were usually sold there. In this particular case, the market activation fully focused on schoolchildren (Box 6.5).

Box 6.5 Business Champions – Snapshot: Banda Borae, Ghana

Banda Borae Cooperative is one of the more grassroots partners with which 2SCALE developed BoP marketing activities. It is a processing group of 20 women, who process and sell soybean kebabs in Kpandai, rural northern Ghana. The cooperative initially marketed their soybean kebabs at the local village market, without a branding or retail strategy.

Sales were seemingly effective, especially on market days in Kpandai, when all the other villages came together to buy and sell goods. However, sales remained stable and little growth was achieved.

Together with 2SCALE, the cooperative managed to expand its route to market by introducing a branded container as a sales outlet in a number of schools. This solution came about during the Business Model Canvas workshop and discussion on marketing issues. The strategy was supported by stakeholder engagement with school authorities (school heads and teachers) to directly target the schoolchildren for nutritional purposes.

The partnership also developed a mobile channel (containers) to enable women to sell soybean kebabs in branded containers and attire in various locations to enhance their appeal and visibility, which turned out to be the most effective distribution strategy.

Over the course of a short six-month pilot period, these two marketing channels increased sales of the kebabs on a daily basis, from an initial average of 150 sticks to 500 sticks per retailer. The cost of a stick of kebab is 10 Ghana pesewas (US $0.02) and provides a nutritious, protein-rich alternative to the sugary snacks schoolchildren would otherwise purchase.

During market activations, the product is often introduced in a playful and fun way, for instance, in combination with dancing, singing songs, and playing music. MCs ensure visitors are enthusiastic about what is happening and create a buzz. Next to the physical location where the activation is held, vans or other motorized vehicles tour the neighborhood to gather a larger crowd to the location of the market activation. In Ethiopia, during the market activations with business champion Family Milk, as part of the larger dairy partnership, locally celebrated singers were involved in the market activation. In Nigeria, a market activation for a Tom Brown product in Kaduna state greatly benefited from the visit of the governor's wife to the activation activities. And obviously, a critical element in all market activations is to provide consumers the opportunity to taste the product.

Lastly, coupons or other actions are sometimes introduced to attract consumers during the activation to buy the product at a discount (the Action in ATEAR). These market activations are also the right moments to inform potential BoP consumers about the importance of nutrition. In the aforementioned pilot on soy kebabs, activities were developed, together with school teachers, to educate the children on the importance of a healthy diet and how that could improve their performance both at home and in school. Combining the general benefit of such activities, the schoolchildren also better understand why it is more beneficial for them to buy a protein-rich soy product instead of sugary snacks; this will help drive demand for the soy kebab sticks.

6.4.6.4 Last-Mile Distribution

Even when a product is available and properly marketed, there are no guarantees that the product will actually reach BoP consumers. This was clearly the case in the first pilot 2SCALE delivered with GUTS Agro, where together we developed the Supermom product completely from scratch and ran an intensive marketing campaign to promote the product. However, the product uptake was less than expected, and distribution was a larger than expected part of the challenge.

As mentioned earlier, BoP consumers are typically removed from conventional market outlets. This means that sales points have to be located closer to the consumer, or distribution should be organized in such a way that it directly reaches the consumer. Previously we referred to three strategies through which such "last-mile distribution" can be organized in a way that does not become too costly and the end product too expensive.[11]

- *Use of existing channels and infrastructure:* The first and easiest approach is to use existing distribution channels, also referred to as "piggybacking." To explore such approaches, the business champion needs to investigate the existing channels in the BoP target market. For example, there could already be a distributor active in the market that sells other, non-competing FMCGs such as bottled water. If this distributor has sufficient reach, the business champion could explore a potential collaboration. The advantage is that such an approach may reduce the business champion's own investment in setting up a distribution system. Also, this can be done quite quickly and easily. The disadvantage is that there is limited control. Therefore, it is often suggested that business champions build initial sales themselves before involving external distributors.

- *Hybrid partnerships:* The second approach is to explore hybrid partnerships. The difference from the first strategy is that hybrid partnerships focus on organizations that are not primarily focused on selling FMCGs. These could be nongovernmental organizations (NGOs), microfinance institutions (MFIs), self-help groups (SHGs), or cooperatives. For example, an NGO that works on promoting health messages in a BoP community might also be interested in promoting and selling nutritious porridge. The benefit of such a hybrid partnership is that there might be strong synergies, without any pressing business competition. The downside is that hybrid partnerships can take quite some time to establish, and they can be political in nature. These partnerships are not necessarily driven by business acumen, limiting their commercial viability.
- *Micro-franchise:* The final strategy is the micro-franchise model, in which a business champion works on setting up a network of individual micro-entrepreneurs. These micro-entrepreneurs can function as sales agents, either by having a small shop or pushcart or going door to door. The product portfolio that these agents sell can exclusively belong to the business owner or include other FMCGs. Building such models requires significant investment. All the agents need to be recruited, trained, and provided with marketing materials and stock. After that, it requires thorough management to ensure the agents perform well. The main advantage of this approach is that the business champion has better control of aspects such as pricing, product placement, and more. It can take years before the model is fully operational and has reached scale. It is often a model that has high appeal among business champions; however, it is important to keep these challenges in mind before adopting such an approach.

As mentioned above, GUTS Agro's Supermom food product did not sell as fast as we had expected. So, the business looked at a more direct distribution method that utilized different wholesalers and increased GUTS Agro's control over distribution. Ultimately this led to the setup of the Likie Ladies' network, a network of around 100 independent female sales agents that distribute GUTS Agro's products door to door.

Also, as part of the pineapple partnership in Benin, business champion Promo Fruits set up a micro-franchised network of sales agents, as it wanted to have control over the distribution and sales of its fruit juice. Additionally, it was an opportunity to spark micro-entrepreneurship locally. It was also a good way to access firsthand customer feedback through the agents that interacted with consumers daily. At the same time, the model was relatively cost intensive, so Promo Fruits developed a business plan for the distribution network to secure working capital to tackle cash flow challenges (Box 6.6).

Box 6.6 Business Champions – Snapshot: Promo Fruits, Benin

Back in 2000, a group of pineapple producers, led by Dieudonné Alladjodjo, unsuccessfully tried to develop different trading relations with larger off-takers in Benin, Nigeria, and the European Union. The group decided that processing pineapple juice for the local market was more commercially viable and created the Pineapple Recovery Initiative (IRA) cooperative, with starting capital of US $5,000 and a processing capacity of 200 kg of pineapple per day. The IRA cooperative gradually increased its processing capacity to 5,000 kg per day in 2009.

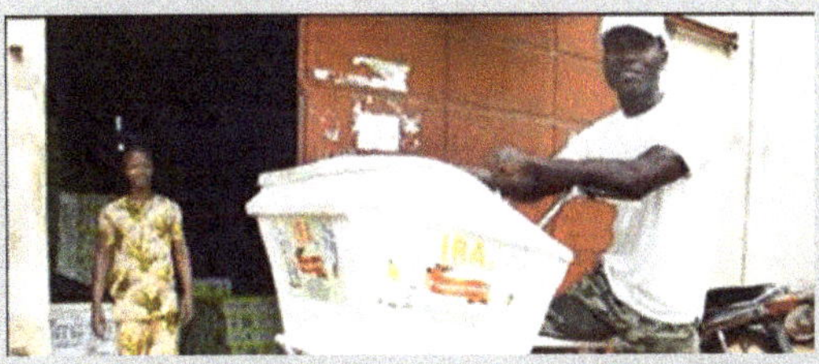

Then in 2011, the IRA cooperative was transformed into a company with limited liability (SARL) and 100% of its shares distributed among IRA member producers. The company's processing capacity increased from 5 to 45 tons per day.

The business model of Promo Fruits is based on four pillars:

1. The company sources from small-scale pineapple producers in Benin, including 2,580 producers grouped into nine professional organizations.
2. Promo Fruits produces 100% natural juice, for which the company has been able to develop a national and sub-regional market.
3. The company offers producers a competitive and incentivizing price, which motivates and retains producers and allows the factory to guarantee its supply.
4. Promo Fruits intervenes upstream in the value chain by facilitating the access of producers, who are members of the sourcing network, to an input credit system.

Over the course of collaboration between Promo Fruits and 2SCALE, the company's management indicated the intent to focus more specifically on BoP markets. Without changing the nature of the product (natural fruit juice), new packaging was developed. Smaller SKUs were produced to increase affordability, and a micro-franchise model was tested; initially, there were 15 sales agents, (five on cargo bikes, five with pushcarts, and five on foot, going door to door), all selling not only the juice but also making combinations with sandwiches. The agents performed well, and at the end of the collaboration between 2SCALE and Promo Fruits, a business model was written to attract funding to sustain the network independently.

In Ghana, via the soybean partnership, business champion Yedent Agro took another approach by piggybacking off the existing networks of "koko sellers"[12] – through which the company decided to distribute and sell its Maisoy Forte Tom Brown mix. This was a good channel for Yedent Agro to expand its exposure toward BoP consumers. However, this strategy limited the company's influence over the final sales price of the product, which was determined by the koko sellers themselves. This became a beneficial strategy for Yedent Agro. They first wanted to test this new model on a smaller scale and with limited resources before considering scaling up. However, the first test with the koko sellers was so successful that the company decided to continue with the model and scale it up.

6.5 Lessons Learned

During the program, the 2SCALE team collected different lessons learned, as this rather pioneering work with African food-producing SMEs and farmer organizations did not always run smoothly. The following are a few lessons learned that could help strengthen the potential for success of any programmatic approach to BoP marketing.

6.5.1 The Opportunity Is Real

In the beginning of the design and implementation of the 2SCALE program, it was not foreseen that the BoP marketing pilots would become a key intervention in several of the partnerships. During the program, the value of the BoP marketing pilots became apparent to the project partners and the business champions. Especially when the first successful cases were established, other business champions were motivated to follow a similar path. This development has shown us that the pilots not only offer value for the program and champions involved but also that there is a real market opportunity in terms of serving the BoP with nutritious quality foods. BoP consumers have proven to be open to try, use, and repeatedly purchase the products that were launched by the champions. So yes, the opportunity is real – from yoghurt to porridge and soy drinks! For all pilots that were implemented, the majority are still running and serving the BoP. This is the reason that for the second phase of the 2SCALE program, the official target is to reach one million consumers, necessitating that 2SCALE continues to support more business champions accessing local BoP markets and serving them with nutritious food products.

---

[4] Koko is a common type of warm porridge that is produced on the spot by street vendors called "koko sellers." They run their preparation and sales out of simple stalls on the roadside across Ghana.

6.5.2 Consumer-Centered Approach Is New for Most Business Champions

As explained in this paper, one of the first steps in building a pilot is to develop an in-depth understanding of the market and the consumers. This consumer-centered approach is sometimes considered as a default way of approaching new business development. However, for most business champions in the 2SCALE program, this was a novel and new approach. For some champions, it was a struggle to have to first invest in research before being able to take action. Even though the business champion could target its new products at local communities next to the factory, it was still a big step to actually go out in the market and gain a deep understanding of the end consumer. Ultimately, all successful champions did embrace this more lean, consumer-centered approach. In the 2SCALE program, it has therefore been a critical step in all BoP marketing pilots and will be included in future partnership development.

6.5.3 Actionable Insights as Catalyst for Progress

At the start of the 2SCALE program, significant time and resources were invested in conducting general market research on the specific 2SCALE sectors in the various countries. However, it became clear that this general market data provided little use for the business champions. In short, market data is not the limiting factor for champions to develop products and marketing strategies. Instead, business champions search for tangible product and market opportunities, applying a lean approach in leveraging these opportunities. During the course of the program, 2SCALE tailored its approach by conducting smaller market/consumer research activities that were tailored to the specific business champions. This resulted in actionable insights that the business champions actually could use, such as a practical way of mapping local competitors, pricing, and understanding consumer desires and behavior. Once it became apparent that these practical and actionable insights were useful for the champions, this research approach was further integrated across the program.

6.5.4 Transformation of the Business Champion Is Needed

Particularly for business champions that have a business that is not (yet) dealing directly with end consumers but rather with other businesses or governments, BoP marketing (or any other direct consumer marketing) is a radically different type of business. Instead of dealing with large volumes, contracts, and buyers, the BoP market progresses in a slower and more gradual manner. In addition, a creative approach is required to understand consumer needs and desires and, in turn, build relevant products and marketing strategies. Suddenly CEOs need to make decision on various "creative considerations," such as what label designs, colors, and images are needed. It also requires a different approach in distribution and sales, dealing with many small volumes and a large number of cash transactions. All in all, working in the business-to-consumer (B2C) channels requires a specific set of qualities and skills that many business champions did not have at the start of the partnership. In some cases, the business champion did not manage to make this adjustment and/or build a B2C team. However, those that did succeed made that adjustment and transformed their business approach radically. For instance, GUTS Agro is now working with over 100 sales agents that all require dedicated product supply, day-to-day support, and training. Based on these experiences, the 2SCALE program now aims to prepare business champions in more detail on building successful B2C channels and its related transformation in thinking. Again, the lessons learned from the first business champions have helped tremendously in preparing others. That is why 2SCALE also promotes cross-learning between champions and organized field trips to successful case studies for new or potential business champions. Perhaps most importantly, the business champions are guided through this journey by the 2SCALE team.

6.5.5 Activating the Market Is a Must

Another key learning has been the crucial value of activating the market, from the market product launch to the related promotional activities. Selling a new product to the BoP is not easy and does not happen by itself. Introducing the product into local supermarkets or simply selling it to wholesalers and sales agents is not enough. It requires a clear and specific strategy, including which distribution channels will be used (e.g., women groups) and which marketing activities will be organized (e.g., tasting sessions) in order to accelerate product sales. This approach is explained in detail in this paper, and it has been proven repeatedly as a fundamental step to help kick start a business. At the same time, both the business champion and 2SCALE have to be cautious of the relatively high cost of such market activations and make clear arrangements on how these costs can be shared to be as equitable as possible. Now that there are several successful case stories on market activation, it becomes easier for business champions to replicate certain approaches. Cross-learning between champions is therefore facilitated in the program as well as promoting learning and results via blogs and videos.

6.5.6 Getting Stuck by Missing or Broken Machinery

As mentioned in Sect. 6.2 under conducting a needs assessment, 2SCALE learned that it is critical for the success of a pilot that a business champion has the right machinery in place and is ready for production, especially when dealing with larger business champions that produce processed and/or packaged foods and rely heavily on machinery to produce their food products. In several cases, pilots were heavily delayed or did not progress further than the strategy phase, because the product simply could not be produced due to the delayed procurement of the right machinery (or the right volumes of raw materials) or lack of local technicians available to service or fix machinery that broke down or did not function properly. This makes all the preparatory work a waste of time and does not fit the context of piloting, where approaches and strategies will need to be tested under a short time span. Doing a proper assessment with the business champion on their "capacity to produce" is paramount, and one should be cautious with promises regarding bank loans to be secured, grants to be received, or technical support to be available.

6.5.7 Adhering to Food Safety Regulation Is of Essence

In many of the pilots, the strategizing activities eventually led the business champion to come up with a completely new food product. However, if a company comes along with a new product, especially when the company has finished "market testing" their new product and wants to launch the product in a larger market, the new product often will need to be tested by the food safety authorities and receive a certification that it meets food safety requirements. In some instances, these processes have proven to be very time and resource intensive; particularly with the smaller business champions, management did not always have the right expertise or capacities in place to deliver on the different requirements or requests for information that these government agencies ask for. Therefore, in different instances, as part of an exit strategy, 2SCALE has supported the business champion in meeting the requirements to get their product approved and certified by food safety agencies.

6.5.8 Preventing Adverse Environmental Effects

In several of the BoP marketing pilots, new products were developed, or new stock keeping units (SKUs) of products were introduced. Often, these developments also cause larger volumes of packaging materials, often plastic, to be introduced to the market. When not addressed properly, this increase in packaging materials could lead to substantially more plastics to be dumped as garbage and affect the environment. This challenge is not just specific to 2SCALE and touches many different market-led initiatives focusing on BoP markets. Large multinationals like Unilever acknowledge the challenge and are looking at alternative, often biodegradable, packaging options. In the future, 2SCALE will look at opportunities for introducing more environmentally friendly packaging options or working with companies on recycling of their own waste streams, bearing in mind at the same time that this should not add prohibitive additional costs to the business model.

6.5.9 *Measure the Impact of Activities on a Consumer Level*

It was mentioned earlier that at the beginning of 2SCALE, it was not foreseen for the program to undertake BoP market activities beyond doing market research and sharing outcomes with business champions. When 2SCALE started to engage more actively with business champions in BoP marketing pilots and implemented BoP marketing activities, no system was put in place to measure their impact due to the piloting nature of the activities. In many of the pilots, sales figures and other general business information on the pilot's performance was shared by the company. However, there was little understanding to what extent the pilots actually contributed to a nutritious and/or more affordable diet for BoP consumers. The understanding of the impact at the consumer level remained very generic and indirect. For example, in Benin, we brought a soy-based product on the market that could substitute white meat (containing around the same protein level as the meat) at a 30% lower price, making nutritious food available at more affordable prices, an assumed positive consumer impact. As 2SCALE is a program that strives to make a developmental impact, it is critical that in the future, the program implements better systems to measure the effect of BoP marketing activities at a consumer level.

6.6 Looking Ahead

Since January 2019, a second phase of the 2SCALE program commenced, offering new opportunities to take a more extensive approach toward BoP markets and nutritious food products. This translates into an additional program target that is different from the first phase, namely, a target focused on consumers. The second phase aims to improve affordable access to a nutritious diet for one million BoP consumers. To reach this goal, the lessons learned described in this section will be used to create a more rigorous approach to BoP marketing. There are several approaches that 2SCALE will take to create a bigger impact for BoP consumer markets, in comparison to the impact achieved during the earlier phase.

First, the results of BoP marketing pilots from the first phase will be used to showcase to new business champions, helping to clearly demonstrate that the market opportunity at the BoP is real. We expect that the success of business champions – Shalem Investments in Kenya, Promo Fruits in Benin, Yedent in Ghana, and the more grassroots business champions, such as the Danaya cooperative producing attieke in Mali or the Funtua women's group producing fura in Nigeria – will inspire other food-processing SMEs and producer organizations to do the same. Well-documented case studies will be used to create this ripple effect. We will do this not only through face-to-face encounters, but also through digital learning, such as webinars and podcasts – all building on the experience and insights from the first phase.

Another mechanism for increased impact is replication. Overall, the second phase will have a stronger focus on sector transformation, ensuring that the impact of 2SCALE goes beyond individual partnerships and transforms entire sectors. For instance, building BoP market propositions around the processing of soybeans in Nigeria should not only positively impact the business champion with whom this work is undertaken, but ideally it should impact the entire soybean production sector in Nigeria and even beyond, inspiring like-minded entrepreneurs in other countries where 2SCALE is active.

There will also be more room in the program to work on BoP marketing. During the first phase of 2SCALE, BoP marketing was delivered in a limited, pilot format – working with short timeframes and minimal budgets. Marketing pilots were also undertaken in less than half of the 53 partnerships that were mobilized. In the second phase of 2SCALE, the target is to have a BoP marketing component in at least 40 of the anticipated 60 partnerships and to have at least five of these 60 partnerships primarily focused on BoP markets. For this, the in-country expertise on BoP marketing across all eight 2SCALE countries will be expanded.

The focus also will increase on nutrition. The program will further involve nutrition experts to help assess the nutritional value of the different products developed under the BoP marketing approaches, to ensure all products developed under BoP marketing activities are more nutritious than the current alternatives in the market. From the demand point of view, in close synergy with market activations to be undertaken, 2SCALE will look at the opportunity to develop activities that focus on behavioral change in nutrition to help create more awareness with BoP consumers on the importance of nutrition and to influence their purchasing and consumption behavior for the better when it comes to food products. This will not only have a potentially positive impact on the health of consumers, it will also create demand for the food products that are developed under BoP marketing activities, creating new market opportunities for our business champions.

Finally, the importance of partnerships with other programs, organizations, or companies must not be forgotten here. The new and ambitious target of providing access to a more nutritious and affordable diet for one million BoP consumers will need smart collaborations. This way, 2SCALE can share its lessons learned on BoP marketing and, at the same time, leverage the networks of other organizations and programs that also work on the implementation of business-led approaches to marketing and distributing nutritious food to the BoP.

6.7 Concluding Remarks

As mentioned in the last section, the 2SCALE program is extended with 4 years and will run until 2023. The approach and lessons learned will be leveraged with the agribusinesses that will be part of the second phase of 2SCALE. Apart from that, it is our wish that this chapter will be useful for other agribusiness and incubator programs. We hope you find those business cases inspiring and the lessons learned useful as we all work toward a more sustainable and viable food system in low-income markets.

References

Osterwalder A, Pigneur Y (2010) Business model generation: a handbook for visionaries, game changers, and challengers, vol 288. Wiley, Hoboken

Prahalad CK (2009) The fortune at the bottom of the pyramid: eradicating poverty through profits. Pearson Education Publishers, United States. 358 p

7

Hunger Relief and Innovation

David Donnan

7.1 Introduction

Our ability to feed a growing population has been the topic of government policymakers, economists, NGOs, and most obviously the agriculture community for centuries. In this chapter, we explore the history of hunger relief efforts from communities, religious organizations, and governments and now toward private corporations and innovative agriculture startups. While global levels of hunger and malnutrition have dropped significantly in the last few decades, there are concerns that agricultural advancements and R&D spending are not enough to unlock the true agricultural potential of the millions of smallholder farmers in rural and developing nations. Government spending on agricultural R&D has decreased in the last decade and many of the large-scale NGO activities are now focusing on environmental and sustainability issues. A renewed focus by private corporations' CSR initiatives, coupled with NGOs and startup entrepreneurs, is offering a new approach to innovations in hunger relief.

7.2 Global Issues of Hunger and Malnutrition

Throughout history, agriculture innovation and hunger relief have always been closely connected. The advent of perennial agriculture led to the first permanent settlements over 10,000 years ago, creating the first reliable source of food (Diamond 1998). Agriculture supplied a more stable source of food and nutrients to the early hunter-gatherers and provided the foundation of tribal governance, work rules, tool development, and culture for early civilizations in the Middle East, Asia, and Africa.

D. Donnan (✉)
Kearney, Chicago, IL, USA
e-mail: Dave.Donnan@kearney.com

Structured, sustainable agriculture allowed civilization to take root and develop. Small tribal encampments became villages, then city-states, and eventually nations (Harari 2015).

The development of agriculture and subsequent agricultural revolutions allowed for accelerated population growth. Eventually it evolved from simply enabling population growth to reducing hunger resulting from population growth. As population centers grew, fewer people were involved in agriculture and farmers role shifted from self-sufficient harvests for their families to feeding all the members of the community. How to feed a growing planet has been an issue of public discourse for centuries, perhaps since Thomas Malthus wrote in 1798 that "the power of population is so superior to the power of the earth to produce subsistence for man, that premature death must in some shape or other visit the human race" (Malthus 1798). Fortunately Malthus's warning has not come to pass, thanks largely to various technological innovations (first mechanized farming tools, later nitrogen fertilizers, and subsequently the genetic improvement of crops which through the green revolution, the development of transgenic crops and other scientific progress has enabled dramatic increases of productivity in most crops) that have aided the "power of the earth" to meet the demands of a population that has increased eightfold since the eighteenth century. Recently, the issue of food security has come back into the spotlight (Fig. 7.1). After decades of relative stability, food prices have become more volatile, dramatically illustrated by the 2007–2008 global food crisis that saw riots in over 25 countries.[1]

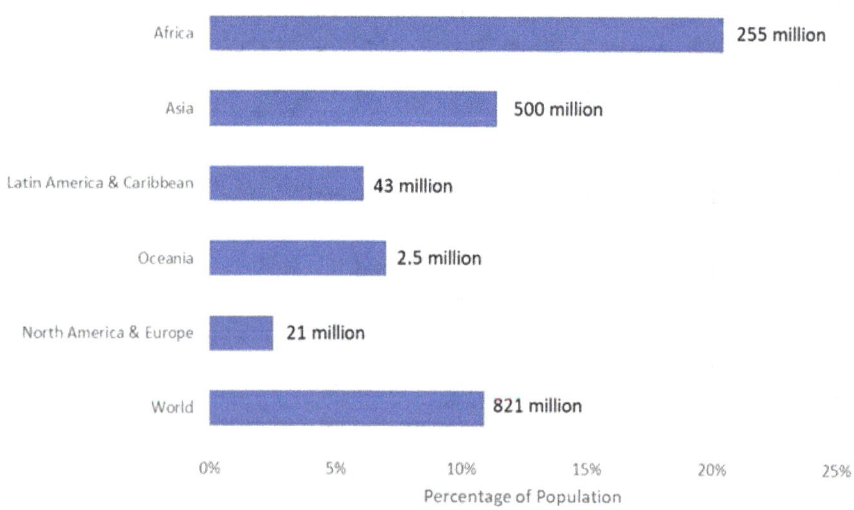

Fig. 7.1 Prevalence of undernourished in the world (2017). (Source: FAO 2018 The State of Food Security and Nutrition in the World. DC Donnan analysis)

Today (2018), hunger has been dramatically reduced around the world. According to FAO "State of Food Security and Nutrition 2018," out of a world population of well over 7.6 billion, only approximately 821 million are undernourished (FAO, IFAD, UNICEF, WFP and WHO 2018). This is a dramatic decrease from 30 years ago when over a 1 billion people were considered hungry and malnourished worldwide. Even with a slight increase in global undernourished population in recent years, hunger has been reduced from 20% of the world population to 11% today (Fig. 7.2). If we look deeper, we find that most hunger and malnourishment exist in areas where access to nutritious food is limited for a variety of reasons. External forces such as drought, insect damage, transportation availability, and geopolitical conflict are much greater influencers of food access than agricultural productivity. We currently are producing enough food to feed the world; it is just not evenly distributed.

That still leaves us with the question of how to feed the future population as we continue to witness continued, although slowing, global population growth. Current estimates predict that the global population will increase to over 10 billion people by 2050, an increase of 2.5 billion or 35%. There is no linear relationship between population growth and the necessary increase in food calories. Resources like land, water, soil are finite. But we can improve their effectiveness by increasing our agricultural productivity or yield (Figs. 7.3 and 7.4). The unknown variables are human diet, climate, and access.

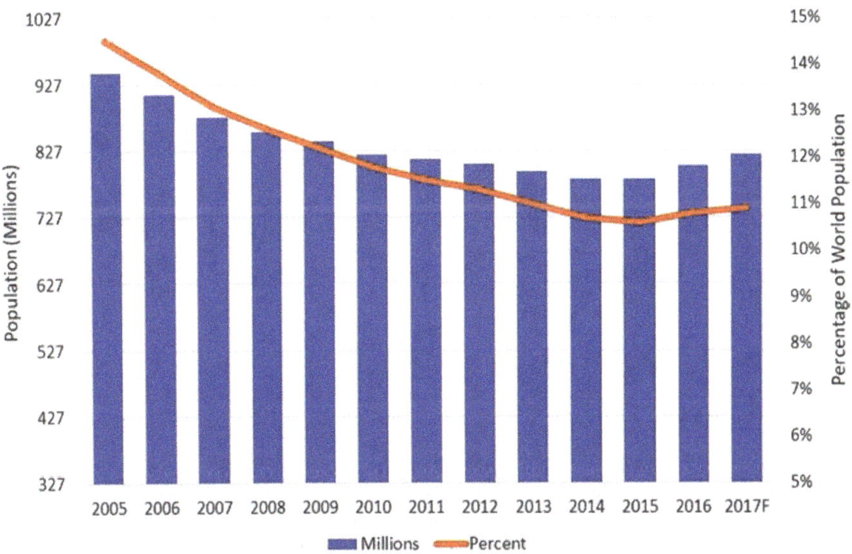

Fig. 7.2 Number of malnourished people in the world (millions and %). (Source: FAO 2018 The State of Food Security and Nutrition in the World. DC Donnan analysis)

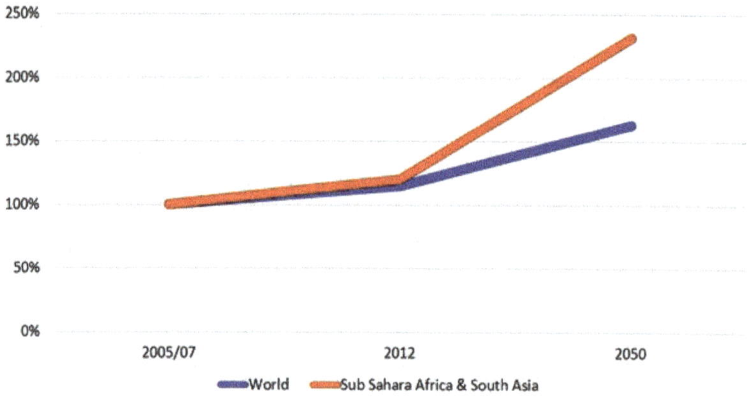

Fig. 7.3 Increase in agricultural output required (to meet projected demand). (Source: The future of food and agriculture, FAO Rome 2017 DC Donnan analysis)

Fig. 7.4 Growth in land and labor productivity (1961–2005). (Source: OECD Library Global and US Trends in Agricultural R&D in a Global Food Security Setting 2012 DC Donnan analysis)

As population groups become more affluent, they eat more processed food and increase their protein consumption to replace plant-produced proteins. As a result, we find the 35% increase in population will require almost 100% increase in crop production to meet the shifting diet and nutritional needs.[2] Climate has always had a significant impact on agriculture from regional droughts to early frosts and excessive rains destroying entire harvests. Access is still the most significant contributor to famine and hunger. Globally we produce enough food but due to lack of infrastructure, geopolitical conflicts, and spoilage, we are unable to efficiently get it to those in need.[2] Finally, as a society, we do not utilize our food resources well. Food waste is a significant issue. It is estimated that between 25% and 30% of all harvested food is wasted either through cultivation, transportation, processing, and consumer plate waste. Managing and reducing food waste can go a long way to closing the food gap between agricultural production and population consumption.[2]

7.3 The Long History of Hunger Relief

Charity and feeding the hungry have always been part of human society. It was first referenced in biblical times in both Christian and Jewish scriptures.[3] Throughout history hunger relief was a localized event focused on the less fortunate members of a community. As civilization evolved, the task of feeding the poor migrated from communities to churches and religious organizations and then to larger-scale government institutions. In the Ottoman Empire, the development of public soup kitchens – which handed out free food to the needy in a combination of patronage, hospitality, and charity – and food banks evolved in collaboration with the churches (Cohen 2005).

Another example of community outreach on a global scale occurred in 1847, during the height of the Irish Potato Famine; a Choctaw community in Oklahoma, USA, who had been forcibly removed from their ancestral land during what came to be known as "The Trail of Tears," pooled its money to help feed starving families across the Atlantic, raising $170 ($5,000 in today's currency) to donate to the people of Ireland. In 2015, Ireland erected a monument called Kindred Spirits to pay tribute to the Choctaw people's generosity. Irish schools still use this history lesson to teach the spirit of truc selflessness.[4]

After World War I, large-scale, government-state-funded hunger relief became more prevalent as shipments of foodstuffs were organized to feed the needs of refugees and inhabitants of war-torn regions. Federal food assistance programs arose during the Great Depression when excess wheat production was diverted to feed the poor and less fortunate peoples in the United States.[5] In the 1940s, the newly established United Nations (UN) became a leading player in coordinating the global fight against hunger. The UN has three independent agencies that work to promote food security and agricultural development. The first, the Food and Agriculture Organization (FAO), created in 1943, is a specialized food and agriculture agency tasked with eliminating hunger, food insecurity, and malnutrition; reducing rural poverty; enabling more efficient food systems; increasing resilience to disasters;

and making agriculture, forestry, and fisheries more productive. The second, the *World Food Programme* (WFP), established in 1961 as a joint FAO/UN joint venture, provides emergency food aid and humanitarian services. The third agency, the *International Fund for Agricultural Development* (IFAD), formed during the food crisis of the early 1970s, is tasked with providing improved technologies and production methods to the poorest nations of the world (Shaw 2007).

Most hungry and malnourished peoples live in rural areas and earn their livings by working on small-scale farming, fishing, or forestry. Almost 75% of these farms, around 375 million, are smaller than one hectare in size and provide barely enough food for the family to live. Programs that helped rural families in times of crisis were often extended to become permanent relief operations. While assisting in immediate hunger and nutrition efforts, these programs did little to improve the long-term viability and sustainability of the rural poor.

After decades of hunger relief programs, a shift in policy emerged in the 1980s. Leading food donor nations became weary of providing excess foodstuffs to other food deficit nations. It was becoming evident that the wholesale export of charitable grains and foods to developing nations was hurting their ability to develop their own sustainable agriculture economy.[6] A new "market-based" approach started to emerge that emphasized self-sufficiency and building infrastructure to support local farmers and sharecroppers. Food aid shifted from exporting food to providing the necessary funds and expertise to develop agricultural food systems within countries and regions (Fig. 7.5). While governments led this shift, the newly emerging

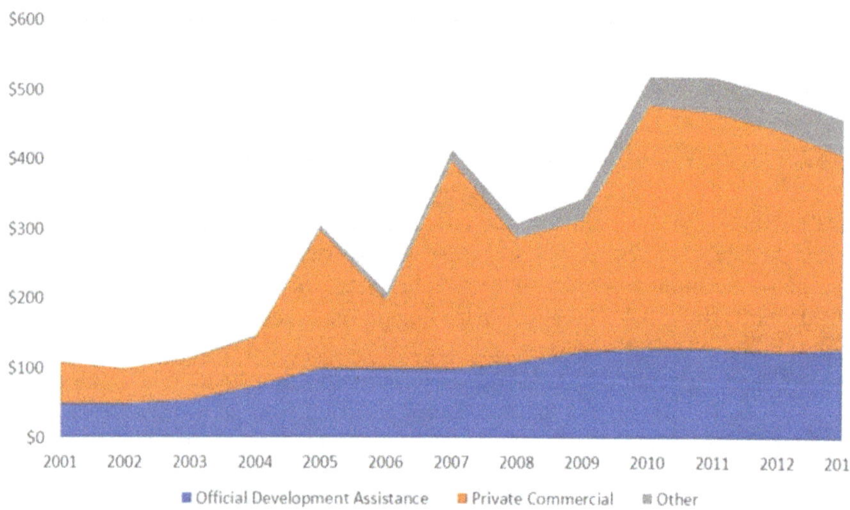

Fig. 7.5 Financial flows to low-income countries (US$ Millions). (Source: The future of food and agriculture, FAO Rome 2017 DC Donnan analysis)

multiregional and global corporations started to become involved since they owned the production and distribution infrastructures need to provide access to food.

7.4 A Shift in Values: The Emergence of Corporate Social Responsibility (CSR)

As governments of developed countries reduced food relief efforts in emerging economies, corporations began to see an opportunity for growth through expanded markets. In lieu of reduced government intervention, corporate social responsibility (CSR) programs began to develop, designed to benefit the triple bottom line of economic, social, and environmental stewardship. The failure of Enron, WorldCom, and Tyco led to an increased mistrust of corporations and their ability to work toward the public good (https://hbr.org/2009/06/rethinking-trust). As a result, governments and NGOs pushed to solidify the need for better corporate oversight and a rethinking of responsibilities to consumers and the society. CSR initiatives are to some degree a result of the need for large corporations to regain public trust and provide better environmental and social stewardship.

CSR initiatives in the food and agriculture sector often started as environmental sustainability efforts to improve all aspects of the supply chain from farming practices to the use of renewable and biodegradable packaging. As consumers became more aware and vocal about environmental issues, large corporations saw the need to communicate positive results in environmental and sustainability activities. Large corporations have embraced CSR programs as they try to shake the image of "Big Business" and become more socially aware. Large agri-food companies also see CSR initiatives as a direct response to more persistent activism by NGOs focused on social justice topics. In a landmark 2011 Harvard Business Review article, Michael Porter and Mark Kramer argued that a new definition of values creation was emerging in the capitalist system (Porter and Kramer 2011). The concept of shared values was focused on a more sustainable capitalism by ensuring that limited resources such as land, energy, environment, and labor were considered in the overall strategy of profit optimization. As they sought to improve their firms' reputations, the resolution of societal problems, once the realm of governments and NGOs, became a focus for large multinational corporations who realized they could create economic value by creating societal value through a more efficient use of resources and reconceiving products and services.

Today, according to a recent KPMG survey, over 93% of the top 250 global companies report CSR initiatives.[7] As they began to value CSR as an indicator of good corporate governance and business growth governments, stock exchanges and regulators are helping drive global corporations' participation in CSR initiatives. In

Fig. 7.6 UN sustainable development goals. (Source: United Nations Sustainable Development Goals)

September 2015, the UN developed a list of 17 Sustainable Development Goals (SDG) (Fig. 7.6), including poverty reduction, environmental stewardship, and prosperity through the elimination of hunger, clean water and sanitation, good health and well-being, and improved life on land and below water, detailed in 169 specific outcome goals targeted to be completed by 2030.[8] Over 43% of the global top 250 companies connect their individual CSR initiatives to the UN SDGs which require governments, academia, private companies, and NGOs to work collaboratively in new and innovative ways.

For food manufacturing companies and retailers, the historic focus on taste and consumption growth has been augmented by a prioritization of consumer nutrition and well-being. Following the 2015 World Health Organization's guidelines on sugar and salt reduction in processed food (World Health Organization 2015), reformulated products were introduced by several large global food companies in 2016.[9] In agriculture, CSR programs have expanded beyond a focus on environmental sustainability to encompass social impact through reduction of slave labor, small farmer economic improvements, rural development, and the health of migrant farmworkers. Leading food and agriculture companies embrace a comprehensive definition of CSR that covers many of their customers' social, economic, and environmental concerns (Box 7.1). As consumers from developed countries became more aware of social injustices in the agricultural business, exporting companies realized the benefits that attention to the living conditions of workers brings to their businesses (Ortega et al. 2016).

Nestlé, www.nestle.com, is a company that has a 150-year history in the food industry. Its global reach and involvement in everything from chocolate to water and nutritional foods have given it the opportunity to work with agricultural operations throughout the world. Nestlé has had a checkered past regarding social responsibility and environment stewardship, and they have been the focus of activist's scorn and intense media scrutiny.[10] Nestlé's response is to increase their commitment to society and nutrition demonstrated through their creating shared value (CSV) program that works on the individual, community, and planet level on programs in sustainability, nutrition, water, and rural development. In their 2017 report, Nestlé outlines one of its initiatives to empower the next generation of farmers. The Farmer Connect Program helps new young farmers to get the necessary training and mentoring to succeed as prosperous farmers. The benefit to Nestlé is to develop a strong farmer base that will provide a reliable, low-cost, high-quality source of supply for their products.[11]

Box 7.1 Launching Collaborative Corporate Social Responsibility Initiatives: The Consumer Goods Forum[1]

The Consumer Goods Forum[12] (CGF) is an international industry association that represents the leading consumer goods retailers and manufacturers globally. With member companies such as Nestlé, Walmart, Coca-Cola, and Procter & Gamble, the CGF strives to drive positive change through and on behalf of its members. The CGF mission is focused on four pillars and seven strategic initiatives. Two of its key strategic pillars are Social and Environmental Sustainability and Health & Wellness.

The CGF provides a global platform for collaborations between members and NGOs, governments, public health authorities, and other relevant actors to discuss areas for program acceleration and change. For many companies, it is difficult to engage with the variety of NGOs, governments, and agencies on a one-to-one basis. Through the CGF initiatives, the member companies can engage with leading NGO organizations such as the World Wildlife Fund, Greenpeace, International Labor Organization, and United Nations Industrial Development Organization. The CGF provides a more unified forum of engagement as well as a strong communication vehicle for its members to show their motivation and persistence toward their CSR goals. Each of the strategic initiatives is led by co-sponsors from the consumer product manufacturing companies and retailers.

(continued)

(continued)

Under the Health & Wellness mission of the association, the members through their board of directors have made joint commitments on behalf of their organizations:

- By 2016: Make company policies public on nutrition and product formulation.
- By 2016: Implement employee health and wellness programs.
- By 2018: Industry-wide implementation of consistent product labeling and consumer information to help consumers make informed choices and usages.
- By 2018: Stop marketing communications to children under 12 for food and beverage products that do not fulfil specific nutrition criteria based on scientific evidence and/or applicable national and international dietary guidelines.

Unlike hunger relief organizations, the CGF is clearly a business-oriented association with the benefit of its members paramount. But they see the clear focus on both employee and consumer health and wellness as a mandate for change. The shift from a charity focus objective to a business and consumer focus benefit has reoriented the nature of the strategies and initiatives. Each company member is free to engage in their own way to develop CSR goals and focus on consumer health through nutrition, fair wages, or safety. This provides a broader approach to the issues of feeding the world through a more holistic approach to well-being, nutrition, and consumer health.

The CGF expects its members to be transparent in both their communications and their reporting. They host webinars, led by members who are looking to share their expertise and best practices on these subjects, and produce resources such as guidelines and toolkits to support members in the implementation of these actions. They also openly discussing them at key industry events like the Sustainable Retail Summit.

[1]Consumer Goods Forum website, www.theconsumergoodsforum.com, and member interviews

7.5 Feeding the World: A Renewed Focus on Innovation

At the same time the CSR movement has brought positive change in global sustainability and social justice, we have seen a steady decline in the funding available for agricultural research – particularly by leading governmental organizations. Throughout the nineteenth century, agricultural research and development expenditures were on the rise as new developments in crop and animal science dramatically improved land yields and labor efficiency. The United States was the largest contributor to agricultural research with the establishment of land grant universities (in the Morrill Acts of 1862 and 1890), state agricultural experiment stations (SAES),

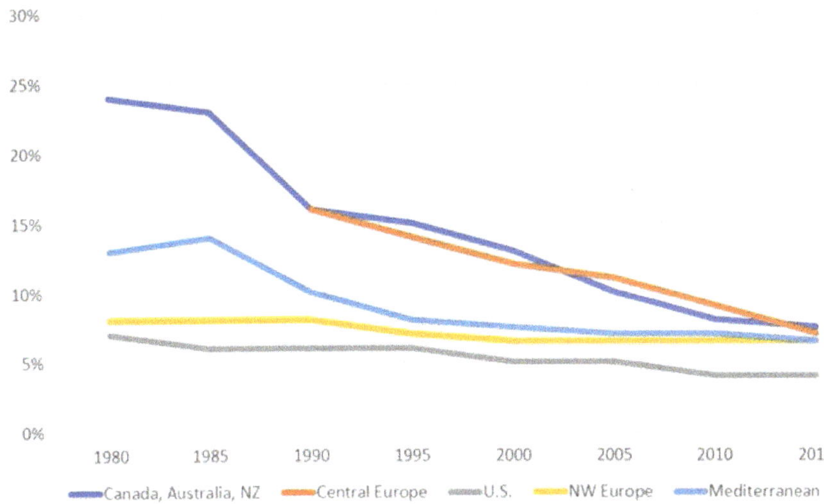

Fig. 7.7 Public agriculture research as a percentage of total public research.

and public-private research partnerships. After the 1960s, many developed nations saw a shift in research to focus more on the environmental impacts of agricultures, food safety, and energy uses of agriculture and away from traditional yield and output improvements (Pardey and Alston 2012).

While government-sponsored R&D has been in decline, there has been a more dramatic increase in private agricultural R&D investments (Fig. 7.7). In 2000, over 55% of agricultural R&D spending in developed countries was through private organizations, an increase of 44% from 1981. In contrast, in developing nations only 6.4% of R&D expenditures are from the private sector with major disparities between regions and countries. The lack of domestic large-scale agricultural companies has been part of the reason for the low proportional investment. But the abundance of small and local farm operations makes large-scale agricultural research and investments more difficult. Most developing nations, apart from Brazil, China, and India, are falling behind in agricultural investments, creating a larger gap between the advanced and developing nations' agricultural capabilities.

Further complicating the reduced investments by developing nations is the nature of agricultural innovation itself. Most improvement technologies used for smallholder farms are either mechanical or chemical applications, both of which tend to be protected by intellectual property rights and patents. This often makes the specific technology application cost prohibitive to smaller farmers in developing nations. Being coupled with the new focus by developed nations away from agricultural productivity toward environmental protection and food safety means that many of the newer innovations may be less applicable to developing economies. This is

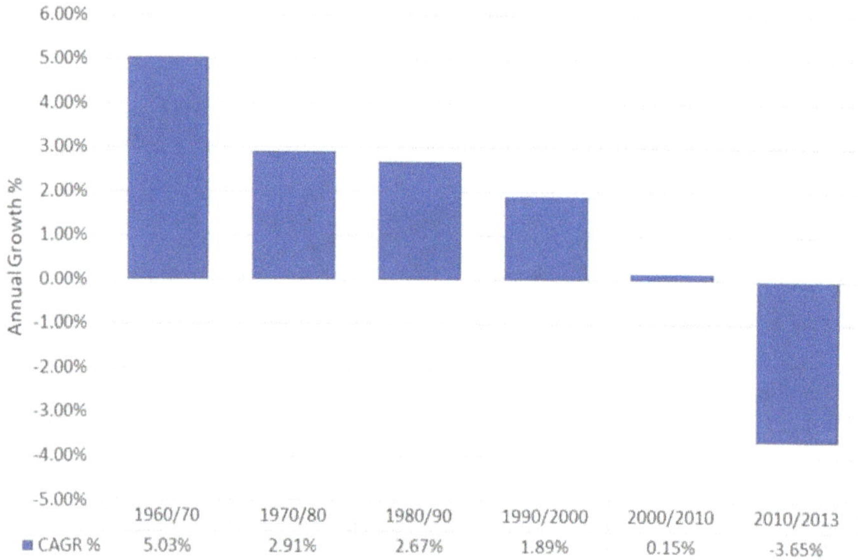

Fig. 7.8 Slowing US public R&D spending for agriculture.

most apparent with larger food and agriculture companies that have adopted the UN SDGs as part of their CSR strategies. Of the 169 targets embedded in the UN SDGs, only 15 are directly related to hunger and poverty eradication.[13] The focus and capital commitment needed to meet the SDGs will redirect funding away from hunger and poverty programs to environmental and sustainable goals such as deforestation, sustainable seafood, and carbon footprint. The gap in agricultural yield and farm output will not be closed with traditional approaches to R&D and investment funding, but will require a new approach to agricultural innovation, particularly for developing nations to close the agricultural productivity gap they experience.

Over the last several decades, the United States, historically the leader in agriculture research, has reduced the growth of its investments in agricultural and crop research, further amplifying the reduction in available agricultural innovation to developing nations and smallholder farmers (Fig. 7.8).

As government investments have been reduced, the private sector has seen an opportunity to invest in profitable research and development projects. Through much of the twentieth century, private equity and venture capital firms invested in new business opportunities and potential growth business operating models. In the late 1990s and early 2000s, a boom in venture funding emerged, foreign investment in the startups during the dotcom boom and Internet bubble. By 2017, venture funds financed close to $150 billion in investments in new technology startups, nearly reaching market highs not seen since the Internet bubble in 2000.[14]

Compared to technology-led industries, the agriculture industry is still underserved by the technology community and venture capital investors. The agricultural community relies mostly on chemical and mechanical technologies for the majority of their innovations in plant yield, productivity, and cost efficiency. Digital technologies did not easily adapt to the rough and dirty environment of farm-level applications. The digital age initially reached agriculture using GPS systems on tractors to aid in seed and fertilizer applications. Precision farming, which allows more precise application of agricultural protection chemicals and nutrients, is now delivering significant value. Technologies such as automation, decision support systems, and agricultural robots are rapidly being adopted on many of the large-scale farms. The connection of smart sensors through the Internet of Things (IoT) gives farmers valuable information, including soil moisture, nutrient levels, the temperature of produce in storage, and the status of farming equipment. Self-driving equipment connects with satellite imagery to apply fertilizer automatically and relay nutrient information to a farmer's mobile computer. Agtech is still small in comparison to other technology investments, but it is disrupting the agri-food industry by applying new innovative and sometimes counterintuitive technologies on the farm and throughout the food processing industry.

In comparison to the burgeoning $84 billion technology-led investments in 2017, the Agtech and FoodTech investment market is still relatively small. According to AgFunder 2017 Investment Report, the food and agriculture venture investments – comprised of over 990 deals and 1,487 unique investors – totaled $10.1 billion.[15] With a 29% increase in investment growth, Agtech deals are getting larger and attracting more prominent investors.

Currently, Agtech investments occur in three broad areas:

1. *Digital Innovation* capitalizes on the latest advancements in hardware and software to create a new system of farming relying on computing power and connectivity. Soil sensors measure ground moisture. Drones collect data and imagery, providing specific and precise crop analytics. Cloud-based advanced analytical solutions process this data to provide growers with direct recommendations. As Randall Barker, Managing Director of FarmLink, states, "There's an obvious convergence about all these technologies that can be connected and solved for in a data science approach. It's all about taking the complexity of biology, weather, and math and making it work."[16]

2. *Biotech Innovation* incorporates scientific techniques to improve plants, animals, and microorganisms. It includes a broad array of solutions, ranging from genetically engineered plants and animals, improved tools such as CRISPR (an improved genome-editing technique with potential in yield improvement, crop protection, ecology, and conservation), and microbial technologies targeting bacteria. Virginia Ursin, Biotechnology Prospecting Lead for Monsanto Company, notes, "18 million farmers globally are growing biotech crops. Ninety percent are in the developing world as smallholder farmers and this technology has made agriculture profitable and has decreased problems with pesticide poisoning. We cannot go back to agricultural practices that will increase the footprint of agriculture. We must decrease the footprint. Sustainable intensification and genetic optimization are essential parts of that."
3. *Process Innovation* introduces new farming techniques that address constraints on farmers' productivity and environmental sustainability (see Box 7.2 for a good example on food preserving). Vertical farming combined with hydroponics/aquaponics allows agriculture to flourish in areas without natural soil. Drip irrigation technology uses sub-surface low-pressure piping to deliver water directly to crop roots, resulting in both better yields and preservation of water resources. Desalination removes salts and minerals from saline water for freshwater uses. Efficient technologies, such as drip irrigation and desalination, are some of the tools that Israel has used to produce a surplus of 20% more water than it consumes. Aaron Mandel, Co-founder and Chairman, WaterFX and HydroRevolution, believes that solar desalination can create an affordable, sustainable water source in water-scarce regions. "We need to focus on redesigning the entire sustainable water model from the ground up," Aaron notes. "WaterFX focuses on leveraging what is happening in the energy industry and applying it

Box 7.2 Connecting Farmers with New Preservation Technologies: HERE Foods[1]

The farm to label market connects local farmers with brands and value-added products beyond the basic agricultural commodities. Framers are always looking for ways to promote their crops and minimize the seasonal volatility that is inherent in commodity products. One example of this relationship is HERE.CO, a Chicago-based company that makes juices, spreads, and salad dressing with ingredients sourced from local farmers. Founded in 2017 by entrepreneurs Nate Laurel and Megan Klein, the company has developed a network of over a dozen indoor and outdoor farms in the Chicago and Midwest US market to provide various fruits and vegetables to its processing plant. The products are sold in over 400 retailers in the Midwest markets under the HERE brand.

What makes this business unique it that it combines the use of a local network of farmers with advanced food preservation technologies. High-pressure pasteurization (HPP) is a cold pasteurization technique that is applied to

(continued)

(continued)
finished packages to kill yeast, mold, and bacteria. Once the juice or spread is packaged, it is placed in a hydrostatic pressure vessel and subjected to over 87,000 pounds per square inch of pressure.

The HERE model works as follows:

- Farmers: HERE has direct relationships with local farmers. As a result, they earn more income, gain exposure, and enjoy predictable price and quantity parameters. By developing unique and exclusive supply arrangements with the farmers, the company can receive very fresh ingredients, reliably.
- Community: The HERE model decreases food miles, using less resources. With each purchase, money gets reinvested in the consumer's local community.
- Retailers: The locally sourced products deliver a unique value proposition to retailers, expanding local products beyond the produce department.
- Consumer: HPP products are unique tasting (very fresh and intense flavors), fresh (local ingredients), healthy (clean label and plant based), convenient, and trusted. By applying scale and vertical integration, they can do this at a competitive price point.

By building a community of farmer/suppliers and providing a value-added approach to local produce, HERE hopes to be able to replicate their Midwest model at other locations in the United States and potentially abroad.

to water. Water is just a form of energy. If we have a sustainable, affordable, scalable source of energy, we can actually produce as much water as we need."

While the new startup entrepreneurs in Agtech/FoodTech show promise, they are still only scratching the surface on the global issues of technology access and feeding those in need. Most startups focus on a commercially viable business model while investors are targeting a profitable exit. Better collaboration is needed between the small startups, venture funds, governments, and corporations to develop innovative, sustainable, and most importantly scalable solutions that can be applied broadly in a global environment. Of concern are the 2.5 billion people that live on the 500 million small farm households in most of the developing economies. Their ability to access innovative agricultural and food solutions is critical to the objective of feeding the planet.[17] For many of these smallholder farms, access to water, improved crop yields, access to capital, and easier delivery to markets are critical resources. Several startups and agriculture innovation have started tackling these perplexing issues:

- *Nicaragua*: The 22 vegetable farmers making up the Tomatoya-Chaguite Grande co-operative worked with TechnoServe to improve access to markets for their vegetables and sell directly to supermarket chains.[18] They introduced drip irrigation and staggered their plantings to increase the number of harvests per year.

- *USA*: Robert Egger founded the DC Central Kitchen and L.A. Kitchens as culinary teaching, waste food reclamation, and feeding centers for low-income citizens. The combination allows L.A. Kitchens to train low-income people in kitchen skills often while using wasted and discarded foods. Working with nonprofit AARP and the Los Angeles City, L.A. Kitchens has flourished under the "buy local" program for several retailers.[19]

- *Vietnam*: Metro Cash & Carry works with local vegetable farmers to improve farm yields and employ better agricultural management including crop rotation and improved fertilizer management. The result is improved quality and food safety for the retailer and improved income for the farmer.

- *Nigeria and Kenya*: Hello Tractor is working with over 22,000 farmers, giving them access to low-cost farm equipment rentals using smartphones to access and rent out equipment. Considered the Uber of farming tractors, the smartphone application connects supply and demand. The app lets tractor owners find other farmers in need and rent out their equipment economically.[20]

- *West Africa*: Easy Solar is a for-profit enterprise with a social mission of making clean energy affordable to off-grid communities in West Africa. It finances easy-to-use, high-quality solar devices to farmers not serviced by the existing electrical grid. Since launching in 2016, Easy Solar has provided electricity to over 40,000 people.[21]

- *Bangladesh*: HarvestPlus is an international NGO focused on biofortification of crops to fight global malnutrition. Most Bangladeshis eat rice, but childhood zinc deficiency and its associated stunting affect 40% of the children. PRAN, a food distribution company, worked with HarvestPlus to supply zinc-fortified rice to farmers and then bought the harvested rice through their distribution system. Over 120,000 participating farmers were able to grow a profitable crop and reduce malnutrition.[22]

- *India/Africa*: Digital Green is a unique partnership among technology companies (Oracle, Google, Microsoft), governments (India and Ethiopia), and NGOs (Bill & Melinda Gates Foundation, USAID). Smallholder farmers are given access to data capture and analytics to improve the overall yields and quality from their farms. Digital Green brings technology startups together with practical implementation opportunities to prove and scale the technology applications across regions and globally.[23]

By aligning incentives based on profits and CSR goals, private businesses are now becoming more involved in localized initiatives to improve farming practices and introduce new technologies to local and small-scale farmers. In some cases (Boxes 7.3, 7.4 and 7.6), they facilitate the access of small farmers to markets and market information and increase their likelihood of commercial success. Technology startups are tackling the issues of hunger and food waste in increased numbers. The entrepreneurial founders of these companies see the elimination of hunger and the improvement of smallholder farmers as being of great benefit – societally and economically. They use their knowledge of crowdsourcing, mobile telephony, and localized initiatives to develop programs that can easily scale using digital technologies. This offers a fundamental new approach to public-private partnerships (PPPs) as the small-scale entrepreneur can tackle larger issues on a smaller scale and then replicate results. In addition to the tech entrepreneurs, private companies and NGOs are working together, creating innovative financing tools and technology transfer mechanisms to meet the needs of the smallholder farmers.

Box 7.3 Connecting Small-Scale Farmers in India: Crofarm and Retail Stores[1]

The farm economy in India accounts for 18% of the GDP and provides employment for 50% of the country workforce. While large, it is also unorganized with small-scale rural farmers selling through local village mandis that procure the produce from farmers and then sell it through a series of middlemen eventually delivered to retailers. This results in a fragmented and disjoint supply chain with wide variations in process and income paid to farmers. While India is the second largest producer of vegetables and fruits in the world, almost 25–30% of the products get wasted due to spoilage from poor transportation and cold storage. For Reliance Retail, India's largest retail enterprise, the fractured supply chain results in excessive spoilage and poor quality of the fruits and vegetables delivered to its stores.

In 2016, two entrepreneurs, Varun Khurana and Prashant Jain, established a business called Crofarm to find a solution to the issues of farmers supply of fresh fruits and vegetables to retailers. They had previously worked for Grofers, an online grocery delivery service that operates in 13 cities in India. There they saw the disjoint nature of the farm to retail supply chain for perishable products such as vegetables, fruits, and exotics. After Grofers they spent time with local farmers in Haryana, Uttarakhand, and Uttar Pradesh to understand the issues faced by the farmers trying to sell their products.

From this knowledge and their understanding of retailer needs, they developed a series of applications to link farmers directly with retailers and bypass

(continued)

(continued)

many of the small village mandis. The applications help farmers select the crops that will fetch the best prices and are in demand as well as connect each farmer to more buyers. By providing better price visibility, the farmers can select the best buyer and time and place to sell. All transactions are digital allowing for quick payment transactions. Crofarm generates revenue through commission, starting from nearly 5% of the price in case of less perishables like potato and onions. It makes a commission of around 15% of the price of green vegetables and 20–25% in case of fruits.

The Crofarm network now connects over 10,000 farmers with leading retailers in India such as Reliance Retail, Grofers, Big Basket, Jubilant FoodWorks, Big Bazaar, and Metro Foods. They have secured cold storage to preserve the fruits and vegetables that require refrigeration. It has also developed an application to connect farmers directly with consumers and utilizes demand prediction algorithms using artificial intelligence and machine learning.

Because of these initiatives and applications, the farmers in the Crofarm network have access to more buyers and better prices. This has helped improve their standard of living while at the same time providing consumers with fresher and higher-quality vegetables and fruits.

[1]With $1M Topline, Crofarm marches toward eliminating middlemen between farm and retail https://entrackr.com/2018/02/crofarm-eliminate-middlemen-farmer/. Crofarm website www.crofarm.com

Box 7.4 Connecting Coffee Farmers in Ethiopia: Ethiopian Commodity Exchange[1]

In 1984 famine struck Ethiopia with extreme shortages of food and grain crops in the north part of the country and resulted in more than 400,000 deaths. In 2002, another famine occurred after bumper agricultural crops in 2000 and 2001 forced prices to record lows. How could this boom-bust scenario be stopped so that the population would not starve and how can both buyers and sellers of crops be protected against drastic price and volume swings.

Prior to the formation of the exchange, buyers and sellers would often meet locally to exchange product for negotiated prices. Often the buyers would not get payments without significant delays and the buyers were always at risk of poor-quality product or shortages in weight.

The Ethiopian Commodity Exchange (ECX) was established in 2008 as a mechanism to bring farmers and buyers together to trade coffee and sesame seeds and assure both timely delivery and payment. The ECX was established

(continued)

(continued)
with seed funds from the Ethiopian government and USAID and was operated as a public-private enterprise. Quickly the ECX was a success as it provided better market information, standardization of products, and standard contracts. The ECX promotes the following services:

1. Market integrity, by guaranteeing the product grade and quantity and operating a system of daily clearing and settling of contracts.
2. Market efficiency by operating a trading system where buyers and sellers can coordinate in a seamless way based on standardized contracts.
3. Market transparency by disseminating market information in real time to all market players.
4. Risk management by offering contracts for future delivery, providing sellers and buyers a way to hedge against price risk.

The information transparency provided the ability to track coffee movement from farm to buyer organization in an efficient manner, promoting better transportation planning and inventory management. Trucks were able to be scheduled, rather than just being available when they happened to show up at a warehouse. This was particularly important since there was a shortage of delivery trucks in Ethiopian and any inefficiencies were costly. All this resulted in lower costs of supply chain as well as improved crop yields from reduced spoilage.

Behind the ECX is a market information system that ties the various trading systems together. This includes electronic tickers at 200 market sites across the country for disseminating information; farmers' access to mobile phones to disseminate market information via text; a fully automated telephone system which allows traders to access market information 24 hours a day, 7 days a week; and the ECX website that provides real-time data for all commodities traded. The rapid progress of technology was manifested by the increased penetration of mobile phone technology in the nation. Between 2008 and 2016, the number of mobile phones increased from under 2 million to over 50 million for a population of 102 million people.

For all the benefits that the ECX proposed, today it is still primarily an export exchange for spot pricing of two primary crops – sesame and coffee. More works need to be done to find a better platform for smallholder farmers to access a larger portion of the overall commodity profit pool.

7.6 How Profit and Nonprofit Can Work Together

Very high levels of investments are needed to use agriculture to solve the issues of malnutrition and rural poverty, but direct investments have been lacking. With limited government resources available, a new model of collaboration among governments, NGOs, and corporations is being sought to improve agricultural productivity and food access. Governments eager to show progress on social programs have encouraged private sector companies through incentives and taxation breaks to enter program investments. Public-private partnerships (PPPs) were utilized in developing countries to solve large infrastructure issues related to the energy and water sectors. In many cases these PPPs were focused on concessions on contracts and leases and independent energy producers. Many communities and NGOs are skeptical about the true intent of PPP initiatives and whether they were just another means to ultimately privatize public infrastructure. In agriculture, PPPs must operate differently than the water or energy sector since the participants are often independent farmers rather than utilities or infrastructure projects. Agricultural PPP programs, while successful, have had to face several hurdles (FAO 2016):

1. Access to specialists in agriculture, land use, and technology. These usually are supplied by the private sector, but for ongoing success, these skills must be transferred to the community.
2. Access to capital and credit. Often this is the role that the government can play or financial institution with micro loans and farmer credit.
3. Poor infrastructure will inhibit the transportation and storage of food items. Basic road and storage systems must be considered before the program can start.
4. Land rights are often not well understood in many developing nations. Land acquisition and land leasing must be part of the program.

PPP contract relationships are more complex in this sector as farming communities are much more fragmented and lack a unified voice, requiring more community involvement and communications.

While many examples of public-private partnerships exist, they are not as simple as they may seem. For private companies to work with NGOs, they need a knowledge transfer framework, financing, resource availability, and the ability to navigate through local regulatory and government rules. Governments must make the process more navigable by redesigning tax incentives, financing rules, regulatory requirements, and access to public resources.

Typically, the roles of the public agency and/or NGO in the PPP programs are as follows:

- Creating a supportive regulatory environment with appropriate incentives for private sector investment and inclusion of smallholders
- Developing program concepts in alignment with national socioeconomic and sector development priorities
- Designing detailed program guidelines and transparent partner selection criteria
- Promoting the incorporation of risk sharing/mitigation in the design process

- Managing evaluation and selection processes for partnership proposals
- Coordinating negotiation and contract signing
- Ensuring regulatory compliance, including the enforcement of land rights
- Providing funding with assistance of larger for-profit companies
- Linking private partners to local public institutions and services
- Providing technical and managerial assistance
- Monitoring and evaluating the partnership at both the national and local government levels

For private businesses, the roles include:

- Developing business plans with thorough financial and market analysis
- Contributing funding or in-kind resources as agreed
- Leading implementation of partnership activities and delivering results
- Providing professional management
- Securing markets for end products and procuring raw materials from farmers through contract farming agreements
- Providing technical assistance and business management training for FOs
- Disseminating inputs and technology
- Linking farmers to business development services (BDSs) such as financing and third-party certification
- Supporting the monitoring of partnership activities

Companies and NGOs have discovered several reasons to partner:

- **Creating business value and environmental benefits.** A business-NGO partnership can result in measurable business and environmental benefits such as reduced costs, reduced risk, new market development, and enhanced brand value along with reduced environmental impacts in the company's product line, operations, or supply chain.
- **Raising the bar on environmental performance.** Innovations arising from partnerships can create competitive advantage for a business as well as establish a new standard of environmental excellence for others to build on.
- **Leveraging skills and perspectives not available in the organization.** Partnering with an NGO can help a company address issues that it may not have the expertise, skills, or resources to manage on its own. NGOs also provide a valuable outside perspective. For the NGO, a partnership can provide a testing ground for the effectiveness of its approach to an issue.
- **Building respect and credibility.** When a partnership between a trusted NGO and a well-known company delivers tangible results, it improves the image and credibility of both organizations.
- **Providing independent validation.** NGO participation can provide independent "third-party" validation of a company's claim of environmental and social benefits from a project.
- **Helping achieve a long-term vision.** While most leading companies and organizations have long-term goals and visions, they often are preoccupied with short-term priorities in their day-to-day operations. A partnership project designed to

address a long-term issue can help provide the external push needed to realize long-term goals.

In some situations, partnerships may not be practical. If the need is immediate or is one that can be achieved independently, a partnership might not make sense. Similarly, an NGO may determine that a partnership is not the best way to achieve a policy goal or social outcome. In general, if a company or NGO can accomplish its goals on its own, there may be no need to partner.[24]

For a private company, a PPP partner will have specific goals in their core areas of focus. By better understanding these goals and the NGO's motivations, a company can make sure that it is both meeting its own objectives and satisfying the needs of its partner. Similarly, an NGO should look for corporate partners that share similar social and environmental values (see Box 7.5). Ultimately companies will need to demonstrate a business case for an initiative and NGOs can help frame the benefits and costs such that the company can achieve both. Many PPPs start out as pilot programs used to demonstrate new technologies or techniques and can be scaled once the benefits are understood.

When working with smallholder farmers, the ownership of the final operating model and business must be transferred to the farmer. This may be difficult to achieve since the ownership also implies risk. Both the NGO and the private company need to be able to mitigate the risk during the pilot phases prior to the scale up of the initiative. This may require additional farmer training, assumption of the initial technology investments, and modifications to the contracts and agreements to allow more flexibility in operations. As the project moves from pilot to scale, the PPP should allow for shifts in conditions to enable creative financing, supply contracts, and inclusion of additional commercial partners. All of these will be required to achieve scale and replicability in other regions.

The Institute of Medicine – Building Public Private Partnerships in Food and Nutrition Workshop is one of the best resources for agriculture-based PPPs (IOM 2012). These text and workshop materials cover the full range of governance and operating rules for companies and NGOs to work together. The workshop covers all aspects of how to choose business partners and how to manage expectations through the initiatives.

Some corporations treat hunger relief and agricultural innovation in ways that do not involve their CSR initiatives. By focusing on profitable economic and social change, the initiatives try to provide an opportunity for seed financing, skills training, and expertise to be transferred to farmers and regional economic systems (Kaplan et al. 2018). This approach is analogous to venture capital and startups. A larger corporation realizes that single-use funding or shared expertise is not enough to develop a sustainable economic and supply system for local farmers. They focus on a well-thought-out business plan, coupled with seed capital and supply chain partners (suppliers and customers) to ensure the sustainability of the new business

Box 7.5 Solving Food Waste and Feeding the Hungry: Feeding America and Starbucks

The concept of food banking was developed by John van Hengel in Phoenix, AZ, in the late 1960s. Van Hengel, a retired businessman, had been volunteering at a soup kitchen trying to find food to serve the hungry. Van Hengel established St. Mary's Food Bank in Phoenix, AZ, as the nation's first food bank. In its initial year, van Hengel and his team of volunteers distributed 275,000 pounds of food to people in need. Word of the food bank's success quickly spread, and states began to take note. By 1977, food banks had been established in 18 cities across the country. Today Feeding America is comprised of over 200 regional food banks that feed over 46 million people in the United States, making it the largest hunger relief agency in the United States.

What started as a small, single store in Seattle's historic Pike Place Market, Starbucks Company has grown into a global leader with 24,000 stores in 70 countries. Starbuck's mission is "to inspire and nurture the human spirit – one person, one cup and one neighborhood at a time." Howard Shultz, the CEO of Starbucks believes that an economic mission must be balanced with a strong social mission as well. "As a business leader, my quest has never been just about winning or making money. It has also been about building a great, enduring company, which has always meant striking a balance between profit and social conscience."

In addition to its coffee and tea products, Starbuck locations sell ready-to-eat foods from their refrigerated counters. Many of these foods are made daily and packaged for a grab and go format. However, at the end of the day, any prepared food that had not sold had to be discarded to maintain the high quality and freshness that Starbucks wanted. Starbucks decided to start a program to donate its ready-to-eat prepared meals from each of its Starbucks stores in the United States to local foodbanks and hunger relief agencies. To be able to scale this effort across the nation, Starbucks decided to work with Feeding America to coordinate nightly pickups of prepared foods and distribute it to the food banks and relief agencies that were working with Feeding America.

The Starbucks FoodShare™ program provides over five million meals to families and individuals served by the Feeding America network. As the program is rolled out to all the company-operated Starbucks stores, the goal is to provide as much as 50 million meals by 2021.

The challenge with the initiative was the coordination of nightly pickups at over 7600 stores and the goal to be in and out of each store in 5 minutes so as not to disrupt the store operations. This required the coordination of hundreds of local food pantries to pick up the packaged foods in refrigerated trucks each day. This allowed the food pantries to pick up sandwiches and other prepared foods nightly and have them ready to be distributed to customers the following morning.

(continued)

(continued)

Jane Maly, Program Manager, Global Social Impact, saw many unanticipated benefits from the program, "In speaking with Food banks, we quickly learned that their concerns primarily revolved around running an overnight shift, seven days per week to pick up food at Starbucks stores. It took a few innovative leaders in the food bank network to address the concerns through a pilot test, which ultimately demonstrated the advantages of operating the program overnight. For example, there are very few vehicles on the road at night, the stores are usually empty of both customers and employees, and the overall logistics of picking up donations are much more efficient."

Based on the success of the FoodShare program and a few passionate people, technology innovators and food startups are now using crowdsourcing and mobile apps to connect restaurants and food donors that have excess fresh food with the food banks, pantries, and agencies in need. Several startups have already been launched across the United States to fulfill this important need.

Source: Feeding America website, www.feedingamerica.org; Starbucks website, www.starbucks.com

Howard Schultz, Joanne Gordon, *Onward: How Starbucks fought for its life without losing its soul*, 2011;

Interview with Jane Maly, Global Social Impact, Starbucks Coffee Company

Box 7.6 Better Life Farming Alliance

Better Life Farming works together with smallholders to help them on their journey to grow their farms into commercially viable and sustainable farming businesses.

Better Life Farming is an alliance of leading private sector companies providing holistic and innovative solutions that enable smallholders to unlock their farming potential. By connecting global expertise with local insights and partners, Better Life Farming offers solutions, each customized to fully address smallholders' needs at scale. Bayer has partnered with the IFC, Netafim, and Swiss Re Corporate Solutions to form the Better Life Farming alliance, specializing in investments in developing areas and providing holistic and innovative solutions that enable smallholders to sustainably produce more food and improve the livelihood of rural communities. Each organization brings their strengths and expertise to the Alliance to deliver the most value possible to smallholder farmers:

Bayer: Bayer is a global enterprise with core competencies in the Life Science fields of health care and agriculture. Its products and services are designed to benefit people and improve their quality of life. Bayer is committed to the principles of sustainable development and to its social and ethical responsibilities as a corporate citizen.

(continued)

(continued)

IFC: A sister organization of the World Bank and member of the World Bank Group, IFC is the largest global development institution focused on the private sector in emerging markets. IFC works with more than 2,000 businesses worldwide, using capital, expertise, and influence to create markets and opportunities in the toughest areas of the world.

Netafim: Netafim is the global leader in precision irrigation for a sustainable future. With 29 subsidiaries, 17 manufacturing plants, and 4,500 employees worldwide, Netafim delivers innovative, tailor-made solutions to millions of farmers, from smallholders to large-scale agricultural producers, in over 110 countries.

Swiss Re: Swiss Re Corporate Solutions provides risk transfer solutions to large- and mid-sized corporations around the world. Its innovative, highly customized products and standard insurance covers help to make businesses more resilient, while its industry-leading claims service provides additional peace of mind. The Food and Agriculture Business team provides a comprehensive array of innovative risk transfer solutions, like crop shortfall covers, weather index, and revenue hedges, to clients along the agricultural supply chain. It also serves farmers directly through corporate agribusinesses and other distribution partners.

By bringing together a group of partners with distinct specialties and skillsets, the BLF alliance provides a comprehensive approach that covers planting seeds, precision irrigation, crop protection, finance, and insurance – all tailored to the specific local and cultural needs of those who farm less than two hectares of land.

Starting in 2016, initial Better Life Farming pilots were launched in areas with high potential for increased productivity in India and Southeast Asia (the Philippines, Thailand, Vietnam, and Indonesia).

India

In Uttar Pradesh, India, the pilot focused on the green chili crop as it is an important cash crop that can provide a consistent income. The project started with 20 farmers in 2016, and since then the BLF alliance has collaborated with more than 1500 smallholder farmers across various villages in the region. In addition, another pilot focusing on smallholder tomato growers was initiated in 2017.

Farmers in India who have joined the program have had tremendous growth, experiencing up to double their previous yields and tripling their incomes.

The Philippines

In the Philippines, the pilot focuses on rice and has sought areas where smallholder rice farmers significantly contributed to the local economy and there was a significant yield improvement potential. The Philippines project started with 40 farmers and 69 hectares of land. From July 2017 to November 2017, that grew to 300 farmers.

(continued)

(continued)

Smallholders who enrolled in the program gained an incremental yield of 57% and a net income 2.7 times higher than the base. BLF aims to partner with 1,767 farmers through 2018, continuing to find ways to increase the agricultural inputs and provide finance for their growth.

venture. As the initiative grows and replicates, a second round of financing is often needing to help capitalize the expansion. In both technology and agriculture start-ups, the same finance and business planning approach is used to inject capital at various stages of development to help progress the initiative and ensure suitable management. This is a longer-term endeavor and requires a multiyear commitment of time and resources.

7.7 Final Remarks

With the evolution of CSR initiatives, coupled with the growth of agricultural start-ups and venture funds, the role of agriculture and hunger relief has dramatically changed in the last several decades. A new focus on sustainable growth, capabilities development, and innovative approaches for smallholder farmers has created an environment of continuous innovation that has replaced many of the donation-based hunger relief initiatives of the past. In the next decade, we should see more socially minded entrepreneurs that will work with NGOs and private businesses to solve the issues of smallholder farming and provide the opportunity to introduce both digital and agricultural innovation at the local level to further the quest to feed the world.

References

Cohen MR (2005) Introduction: poverty and charity in past times. J Interdiscip Hist 35(3):347–360

Diamond J (1998) Guns, germs and steel: a short history of everbody for the last 13000 years. W. W. Norton and Co. New York, USA. Press. Press. 592 p

FAO (2016) Public–private partnerships for agribusiness development – a review of international experiences, by Rankin, M., Gálvez Nogales, E., Santacoloma, P., Mhlanga, N & Rizzo, C. Rome

FAO, IFAD, UNICEF, WFP and WHO (2018) The State of Food Security and Nutrition in the World 2018. Building climate resilience for food security and nutrition. Rome

Harari Y (2015) Sapiens: a brief history of humankind. Random House Press, New York. 512 p

IOM (Institute of Medicine) (2012) Building public–private partnerships in food and nutrition: workshop summary. The National Academies Press, Washington, D.C

Kaplan K, Serafeim G, Tugendhat E (2018) Inclusive growth: profitable strategies for tackling poverty and inequality. Harvard Business Review. pp 365–371

Malthus TR (1798) An essay on the principle of population. J Johnson, London, UK. 272 p

Ortega MI, Sabo S, Aranda Gallegos P, De Zapien JEG, Zapien A, Portillo Abril GE, Rosales C

(2016) Agribusiness, corporate social responsibility, and health of agricultural migrant workers. Front. Public Health 4:54. https://doi.org/10.3389/fpubh.2016.00054
Pardey GP, Alston JM (2012) Global and US trends in agricultural R&D in a global food security setting. In: Improving agricultural knowledge and innovation systems: OECD conference proceedings. OECD Publishing, Paris. https://doi.org/10.1787/9789264167445-4-en
Porter M, Kramer K (2011) Creating shared value. Harvard Business Review January 2011. pp 35–42
Shaw DJ (2007) World food security: a history since 1945. Palgrave McMillan Publishers, New York. 510 p
World Health Organization (2015) Guideline: sugars intake for adults and children, Geneva. 49 p

8

Agricultural Technology, Big Data and Digital Agriculture

Steven T. Sonka

8.1 Introduction

For the future well-being of the global society, agricultural innovation is a necessity. As it has been throughout history, having access to adequate amounts of food is an uncertainty for many individuals across the world. For others, availability of reasonably safe, affordable food is somewhat taken for granted. For both groups, simply continuing the practices of our current agricultural and food system is not sufficient nor tenable as we look to the future.

Among the many stressors facing that system, four are of key importance. Global population growth is expected to continue, particularly in those areas of the world where food security is relatively weak. While this challenge suggests the need for more food to be produced, the environmental effects of agricultural production increasingly are recognized as having both immediate and long-term consequences that are undesirable. Furthermore the prospects of a changing, more variable climate contribute to the need to enhance the resilience of current agricultural practices. And, fourth, consumers in both developed and developing countries are demanding an even more nutritious and safe food supply.

S. T. Sonka (✉)
University of Illinois, Champaign, IL, USA

Ed Snider Center for Enterprise and Markets, University of Maryland, College Park, MD, USA

Centrec Consulting Group LLC, Savoy, IL, USA
e-mail: ssonka@illinois.edu

From one perspective, these challenges are daunting and may seem insurmountable. Yet this setting is not new to mankind and, from a historical perspective, is more the norm rather than the exception. If society is to maintain itself and advance, agricultural innovation is essential. Indeed, 50 years ago, eminent scholars employing sophisticated mathematical models confidently predicted that within a decade massive famine caused by chronic food scarcity would characterize the world's future (Meadows et al. 1972). However agricultural innovation, along with other changes in society, rose to the challenge and led to reductions in the number of malnourished in the world.

Today, new tools such as digital technology and big data are being developed and applied within agricultural production systems. Effective implementation of these tools offers unprecedented capabilities to fuel innovation and contribute to our response to the challenges just noted. (These terms will be described more fully later in this chapter.) It is important to recognize both (1) that their implementation is itself a key form of innovation and (2) that the use of these technologies can foster additional innovation by making existing innovation systems even more effective (Sonka 2016).

While an exciting prospect, the extent and impact of the use of these technologies are themselves uncertain. The purpose of this chapter is to explore that potential for effective implementation. A managerial, not a technological, perspective will be employed as the primary lens for this chapter.[1] A key premise of this perspective is that the existence of a technology does not guarantee immediate or future adoption nor value creation for its users and the market. Rather, the extensive use of technology will hinge on its ability to enable managers to better achieve their goals. These managers can be operating in either the public or private sector or in developed or developing agricultural settings.

Framed by this perspective, this chapter contains the following five sections. First, key elements of digital technology and big data will be described and the most profound decision-making aspect of their application will be identified. This element can be captured by a simple question, "What is, or can be, agricultural data?" While the capabilities of technologies being developed are continually more advanced, the use of what is commonly known as precision agriculture has been in process for the last two decades. Those experiences and lessons learned will be the subject of the second section which follows. The term "big data" has become common throughout much of society, although the term's meaning is not well defined. In next section of this chapter, the characteristics of big data are identified and particular attention is devoted to the central role of analytics. The fourth section of this chapter will depict the digital agriculture that is emerging. Also, that concept will be linked to the broader needs and opportunities associated with the adoption of digital technologies throughout the food system. A brief concluding section ends this chapter.

---

[1] Other perspectives are relevant to the application of digital technologies and big data in agriculture. Space limits preclude their analysis here. For example, Weersink et al. (2018) explore environment implications and Wiseman et al. (2018) consider data ownership and management issues.

8.2 "What Is, or Can Be, Agricultural Data?"

To many, maybe most, of us, the word data tends to generate little excitement. Frankly data was just boring, as it suggested rows and columns of numbers on a spreadsheet offering scant guidance or insight. Yet data in either its implicit (that which we sense or feel) or explicit (that which is written down) forms is essential to how we make decisions. Farmers, throughout time, have been constrained to making decisions based upon what they could observe, sense, and feel. This constraint was imposed by technology and economics. Digital technologies and big data are changing the parameters associated with those constraints. However, those forces will continue to determine the extent and effectiveness of technology adoption.

This section addresses the links of technology, data, and decision making. Its first segment employs a very simple example to demonstrate the interactions of technology, economics, and farmer decision making. The second segment illustrates how emerging technologies are fundamentally changing what is available as explicit data for agricultural decision making. The section's final segment provides a more complete description of some of terms associated with digital technologies.

8.2.1 Measurement

"You can't manage what you don't measure!" is a phrase attributed to both Peter Drucker and W. Edwards Deming. This phrase is as applicable to farmers as it is to managers at Toyota or Amazon (Brynjolfsson and McAfee 2012). The relationship between measurement and the ability to make improved decisions is critically important in understanding the potential for digital technologies to affect agricultural management.

The author of this paper had the benefit of growing up on a small farm in the Midwest region of the United States and, throughout his career, has learned extensively from farmers in the United States and globally. With apologies for a small digression, let me use personal experience to focus on the linkage between measurement and management. Growing up on a farm, the linkage between what could be measured and our ability to improve performance was straightforward. In those days, we had to carry the, hopefully, full milking machine from the cow to the milk tank. The weight of the bucket gave direct evidence as to which cows were producing more. And because there were less than 20 cows in the herd, it also was possible to remember which were the higher-producing cows and give them an extra portion of grain. Laggard producers received less grain.

On this same farm, about 120 egg producing chickens were housed in a building, with ample room to roam outdoors as well. Eggs were collected twice a day. Performance of the entire group was observable. Information that could lead to improved performance of individual birds, however, was not observable. Technically, it would have been possible to establish a production system where measurement of individual bird performance could have been accomplished. However, the economics of egg production and the technologies available at that time did not justify the costs of such a system.

There are two important points illustrated by this story. One is that the desire to link measurement of outcomes and management actions in farming is not new. However, the economics of measurement (the cost of measurement versus the benefits of doing so), given the available technology, inhibited my father and other farmers from capturing and exploiting more data. The second point, then, is that measurement is both an economic and a technical issue for agricultural managers.

8.2.2 *What Is or Can Be Data in Agriculture?*

Suddenly (at least in agricultural measurement terms), the "what is data" question has new answers. Figure 8.1 provides a visual illustration of the change. In its upper left hand corner, we see data as we are used to it – rows and columns of nicely organized, but basically boring, numbers.

The picture in the upper right hand corner is of a pasture in New Zealand. Pasture is the primary source of nutrition for dairy cows in that country and supplemental fertilization throughout the growing season is a necessary and economic practice. The uneven pattern of the forage in that field is measured by a sensor on the fertilizer spreader to regulate how much fertilizer is applied – as the spreader goes across the field. In this situation, uneven forage growth is now data.

Fig. 8.1 Sources of agricultural data. (Graphics courtesy of: agrioptics.co.nz; T. Abdelzaher, Champaign, IL.; Mock, Morrow & Papendieck; International Rice Research Institute)

The lower left hand corner of Fig. 8.1 shows the most versatile sensor in the world – individuals using their cell phone. Particularly for agriculture in developing nations, the cell phone is a phenomenal source of potential change – because of both information sent to those individuals and information they now can provide.

As illustrated in the lower right hand quadrant of Fig. 8.1, satellite imagery can measure temporal changes in reflectivity of plants to provide estimates of growth (RIICE 2013). The picture is focused on rice production in Asia. Such information has numerous potential uses. One is to provide a low-cost means of identifying fields where adverse conditions have caused major production shortfalls. Once that field is identified, similar low-cost means could be used to provide insurance payments to farmers eligible for that insurance.

While satellite imagery is one source of remotely sensed data, recent years have seen a pronounced increase in the capabilities and interest in unmanned aerial systems (UASs) as a source of data for agriculture. There are numerous ongoing efforts to transform UAS technology originally focused on military purposes to applications supporting production agriculture. "Universities already are working with agricultural groups to experiment with different types of unmanned aircraft outfitted with sensors and other technologies to measure and protect crop health" (King 2013). A few, of many, example applications include the following:

- Monitoring of potato production (Oregon State University)
- Targeting pesticide spraying on hillside vineyards (University of California, Davis)
- Mapping areas of nitrogen deficiency (Kansas State University)
- Detecting airborne microbes (Virginia Polytechnic Institute and State University)

8.2.3 Technologies Transforming What Can Be Data?

Terms such as precision agriculture, big data, digital technology, and big data analytics are frequently used in society and among farmers. While such use is common, a common understanding of what these terms precisely mean has not been achieved. (Because of the rapidly evolving nature of the technologies, the problem is not the lack of definitions, rather it is that numerous definitions, all with some validity, exist.)

This section will provide a brief perspective of the terms digital technology, precision agriculture, and big data analytics. The intent is not to provide precise or universal definitions. Instead, the goal is to provide a general perspective that will contribute to a better understanding of the chapter's contents. Further discussion and example applications are included later in this chapter.

The following two-part explanation of digital technology often is useful:

1. Digital technology in agriculture involves:
    - Employing sensors and technologies to capture digital data and operating machines which use digital information to differentially apply inputs

- Using digital tools and techniques to summarize, analyze, synthesize, and communicate digital and other information to improve decision-making

2. Within that broad perspective, it also is useful to distinguish between three types of digital technology application:

   - Precision agriculture: Although having 20+ years of history, precision agriculture technologies continue to markedly improve. Powered by GPS-enabled equipment and machine-based sensors, precision agriculture focuses on measurement and differential input application at sub-field levels. Managerial analysis focuses on the use of data captured from individual farm fields to improve productivity. Over the last two decades, farmers have been exposed to and, in many cases, have had experience with precision agriculture. However, today's advances in sensor capabilities continue to enhance the effectiveness of precision agriculture practices meaning that farmers have an on-going opportunity to choose whether to employ new practices or not.

   - Big data analytics: The ability to cheaply capture extraordinarily large sets of data has fueled numerous big data applications throughout society. However, the existence of massive datasets is only part of the story. Big data analytics requires extensive computational power as well as application of fundamentally different means of analysis to provide probabilistically based insights to improve decision making.

   - A potential for the application of big data analytics in farming is the pooling of production-related data from many farming operations across potentially millions of acres to discern previously unknown managerial insights.
   - Terms such as "big data" and "artificial intelligence" are relatively new to society, let alone agriculture. Tracking of the mention of those terms in media publications (for all uses) indicates that such mentions barely existed in 2007. However, over 150,000 mentions were identified by the year 2014, only 7 years later (Gandomi and Haider 2015). The media hype associated with such rapid growth, however, often contributes to confusion and uncertainty regarding the managerial application of such innovations (Sonka 2015).
   - Some applications of big data analytics in agriculture (weather forecasting, autonomous steering of machines in the field) do not require that detailed farm production data from one field/farm be compared with data from other farm operations.

   - Communication and social media: This category includes two somewhat different applications:

   - The use of social media to communicate with personal and business contacts.
   - The use of telecommunication-based wireless, WIFI, and the Internet

     - To transfer production-related data captured from sensors to devices where that data can be stored/analyzed
     - To transfer findings to the farmer and/or instructions directly to machines as activities that should be conducted

- While typically not featured in discussions of precision agriculture or big data analytics, advances in communication capabilities based upon digital technology often are essential to enhancing performance.

8.3 Precision Agriculture: Precursor to Big Data

This section will provide a brief overview of the precision agriculture experience. It is not intended as comprehensive assessment. It is intended to provide a sense of the evolution of precision agriculture, identify the more popular technologies employed, and discuss the admittedly scant evidence as to the economic gains from the use of these innovations.

It is important to note that precision agriculture and big data are not synonymous. As noted above, the current tools and techniques of precision agriculture have existed largely without the application of big data concepts. However, it is hard to foresee that big data approaches could have significant impact without employing precision agriculture technologies to capture at least of the data required.

Precision agriculture has several dimensions; indeed the concept itself is not precisely defined. A *1997 report* of the National Research Council (National Research Council 1997) refers to precision agriculture, "...as a management strategy that uses information technologies to bring data from multiple sources to bear on decisions associated with crop production." Key technologies and practices included within precision agriculture are as follows:

- Georeferenced information
- Global positioning systems
- Geographic information systems and mapping software
- Yield monitoring and mapping
- Variable-rate input application technologies
- Remote and ground-based sensors
- Crop production modeling and decision support systems
- Electronic communications

The term precision agriculture primarily has been linked to crop production. However, precision practices (and big data techniques for that matter) are equally applicable in animal agriculture, where georeferencing can refer to both sub areas of a field and individual animals. The tracking processes and required tools may differ but the managerial goal is still to separately manage increasingly smaller units of observation.

Farmers and agribusiness managers played a significant role in the development of precision agriculture. For example, in the mid-1990s, a group of agribusiness professionals in Champaign County, Illinois, came together to explore the opportunities associated with two emerging technologies — site-specific agriculture and that strange thing called the Internet (Sonka and Coaldrake 1996). This group, called CCNetAg, was part of an initiative co-sponsored by the local Chamber of

Commerce and the National Center for Supercomputing Applications at the University of Illinois. A voluntary enterprise, CCNetAg provided a vehicle for farmers, agribusiness managers, and university researchers to jointly explore adoption of these tools. The key elements of precision farming are as follows:

- Georeferencing as indicated by satellites linking to the farm field.
- Key farming operations are being directed by and are capturing digital information on the following:
- Soil characteristics
- Nutrient application
- Planting
- Crop scouting
- Harvesting

Since 1997, technologies have advanced, although the general categories remain relevant. For example, auto-steer capabilities on farm machinery have become much more prevalent. And active, detailed measurement of the planting process (recording where "skips" occur) is now feasible. Furthermore, the ability to monitor the status of farm machinery as it operates is now paired with electronic communications to signal when machine operations may be out of acceptable bounds. While there have been many publications describing precision agriculture, reports with independent evaluation of the economics of adoption are much less numerous. One means to assess whether there are net benefits of a technology is to monitor its marketplace adoption. For several years the Center for Food and Agricultural Business at Purdue University and CropLife magazine have surveyed agricultural input suppliers regarding the adoption of precision agriculture. Focused primarily on the Midwest and Southern regions in the United States, this work is a particularly useful assessment of the technology's application. From the *2017 report*, there is clear evidence of adoption for key precision agriculture practices (Erickson et al. 2017).

The crop input dealers who provided input for this study are uniquely well positioned to understand and report on adoption of these technologies. Their firms provide inputs (fertilizer, pesticides, and seeds) and services to producers evaluating and adopting precision agriculture.

Early interest in precision agriculture focused on site-specific application of inputs and on the use of yield monitors. As shown in Fig. 8.3, grid sampling, a practice associated with site-specific lime and fertilizer application, is currently employed on nearly half of the crop acres. Increased coverage to 6 out of 10 acres is expected by 2020. Similar adoption rates have been experienced for GPS-assisted yield monitors. Over the last decade, the use of GPS guidance systems has increased rapidly to a current use estimated to exceed 60%. Continued strong growth to 2020 is expected. The use of satellite imagery and UAVs as tools to support crop production is more recent. Current use affects 19% and 6% of acreages, respectively. Interesting, acreage covered by UAVs is expected to increase by over threefold, to 22%, in just 3 years.

Erickson et al. (2017) describe a relatively consistent adoption pattern for variable rate technology (VRT) practices. In the early 2000s, adoption was at

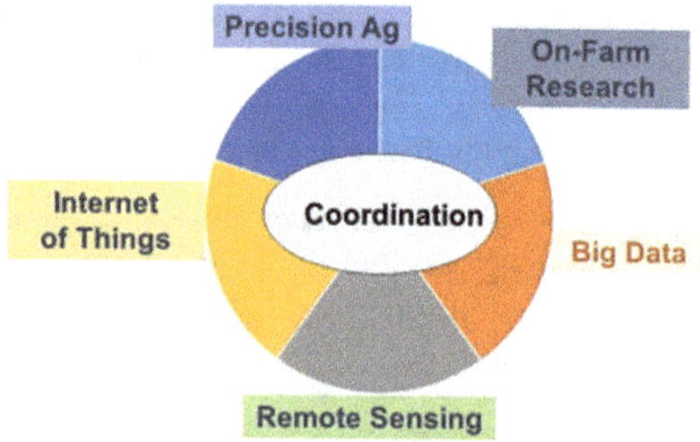

Fig. 8.3 Components of a potential digital agriculture

single-digit levels. Since then, steady increases in the extent of acreage covered have occurred. However, the most utilized practice, application of lime, is only now achieving coverage on 40% of the total acreage. These patterns also are interesting because of the very different price regimes that existed for corn and soybeans over these 15 years. When output prices were low prior to 2008, the driver for adoption likely was cost reduction. Possibly, increasing yields were a more significant factor in later years when prices were higher.

Media and marketing attention sometimes blur distinctions between precision agriculture and big data. Some communications seem to suggest that big data is just an updated buzzword for precision agriculture practices. That is not the case, and the main differences among these two concepts are as follows:

- While the farmer has several types of precision data from each field, additional sources of data naturally reside and originate beyond the fencerow. Accessing that information raises both technical and organizational challenges.
- Precision agriculture employs comparisons across field map layers as its dominant method of analysis. The effect of a single factor, such as a blocked tile line or a buried fencerow, often is observable from a map. However, identifying complex interactions across several production factors and multiple years requires much more sophisticated tools.
- As noted previously, precision agriculture has had 20+ years of experience. Aggregating all the digital information collected from yield monitors and site-specific input operations would result in an extremely large set of data. However, that data currently is located on innumerable thumb drives, disk drives, and desktop computers. Large-scale analysis would not be possible unless/until that data can be accessed and aggregated.

Both precision agriculture and big data arise from the advent and application of information and communication technologies. As noted previously, they are not synonymous. That said, it is hard to foresee that big data approaches will have significant impact without employing the data generated by precision agriculture practices.

8.4 Dimensions of Big Data

Although of relatively recent origin, numerous attempts have been made to define big data. For example:

- The phrase "big data" refers to large, diverse, complex, longitudinal, and/or distributed datasets generated from instruments, sensors, Internet transactions, email, video, click streams, and/or all other digital sources available today and in the future (National Science Foundation 2012).
- Big data shall mean the datasets that could not be perceived, acquired, managed, and processed by traditional IT and software/hardware tools within a tolerable time (Chen et al. 2014)
- Big data is where the data volume, acquisition velocity, or data representation (variety) limits the ability to perform effective analysis using traditional relational approaches or requires the use of significant horizontal scaling for efficient processing (Cooper and Mell 2012).
- Big data is a high-volume, high-velocity, and high-variety information asset that demands cost-effective, innovative forms of information processing for enhanced insight and decision making (Gartner IT Glossary 2012).

Three dimensions (Fig. 8.2) often are employed to describe the big data phenomenon: volume, velocity, and variety (Manyika et al. 2011). Each dimension presents both challenges for data management and opportunities to advance business decision making. These three dimensions focus on the nature of data. However, just having data is insufficient. Analytics is the hidden "secret sauce" of big data. Analytics, discussed later, refers to the increasingly sophisticated means by which useful insights can be fashioned from available data.

"*90% of the data in the world today has been created in the last two years alone*" (IBM 2012). In recent years, statements similar to IBM's observation and its emphasis on volume of data have become increasingly more common. The volume dimension of big data is not defined in specific quantitative terms. Rather, big data refers to datasets whose size is beyond the ability of typical database software tools to capture, store, manage, and analyze. This definition is intentionally subjective; with

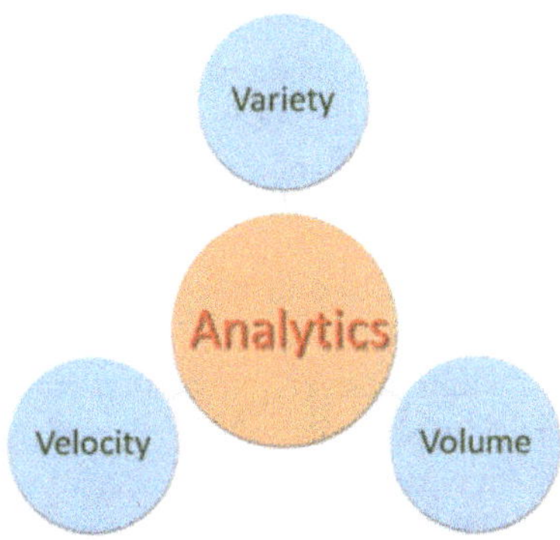

Fig. 8.2 Dimensions of big data

no single standard of how big a dataset needs to be considered big. And that standard can vary between industries and applications.

An example of one firm's use of big data is provided by GE—which now collects 50 million pieces of data from ten million sensors everyday (Hardy 2014). GE installs sensors on turbines to collect information on the "health" of the blades. Typically, one gas turbine can generate 500 gigabytes of data daily. If the use of that data can improve energy efficiency by 1%, GE can help customers save a total of $300 billion (Marr 2014).

The velocity dimension refers to the capability to acquire, understand, and respond to events *as they occur*. Sometimes it is not enough just to know what has happened; rather we want to know what is happening. We have all become familiar with real-time traffic information available at our fingertips. Google Maps provides live traffic information by analyzing the speed of phones using the Google Maps app on the road (Barth 2009). Based on the changing traffic status and extensive analysis of factors that affect congestion, Google Maps can suggest alternative routes in real time to ensure a faster and smoother drive.

For analysts interested in retailing, anticipating the level of sales is important. Brynjolfsson and McAfee (2012) report on an effort to monitor mobile phone traffic to infer how many people were in the parking lots of a key retailer on Black Friday — the start of the holiday shopping season in the United States — as a means to estimate retail sales.

Variety, as a dimension of big data, may be the most novel and intriguing of these three characteristics. For many of us, data referred to numbers meaningfully arranged in rows and columns. For big data, the reality of "what is data" is wildly expanded. The following are just some of the types of data available to be converted into information:

- Financial transactions
- The movement of your eyes as you read this text
- "Turns of a screw" in a manufacturing process
- Tracking of web pages examined by a customer
- Photos of plants
- GPS locations
- Text
- Conversations on cell phones
- Fan speed, temperature, and humidity in a factory producing motorcycles
- Images of plant growth taken from drones or from satellites
- Questions

The variety dimension is closely linked to the discussion of "what is or can be agricultural data?" presented earlier in this chapter. Essentially digital technologies, including those employed in precision agriculture practices, are capturing information as explicit data which previously could only be observed or sensed. Furthermore, in many cases, this process can be accomplished at costs which are economically justifiable. Often times, that newly available data can be directly employed without further analysis. In other instances, the effective use of that information requires the application of newly available analytical approaches – the analytics dimension of big data.

Some agriculturally oriented scholars (Coble et al. 2016; Weersink et al. 2018) will include veracity as a fourth dimension of big data. In this context, veracity references data quality that is employed within big data analyses. Indeed agricultural analysis has a long tradition which emphasizes the need for accurate data, with the oft-used phrase "Garbage in; garbage out," exemplifying this concern. However, as will be discussed in the following section, Analytics, the tools and techniques can produce useful insights from less than perfectly accurate data. Therefore veracity is not included here as a big data dimension.

8.4.1 Analytics

Access to lots of data, generated from diverse sources with minimal lag times, sounds attractive. Managers, however, quickly will ask, "What do I do with all this stuff?" Without similar advances in analytic capabilities, just acquiring more data is unlikely to have significant impact within agriculture.

Analytics and its related, more recent term, data science, are key factors by which big data capabilities can actually contribute to improved performance in the agricultural sector. Data science refers to the study of the generalizable extraction of knowledge from data (Dhar 2013). Tools based upon data science are being developed for implementation in the sector, although these efforts are at their early stages.

The associated concept of analytics similarly is maturing and its use refined (Davenport 2013; Watson 2013). Analytic efforts can be categorized as being of one of three types:

- Descriptive efforts focus on documenting what has occurred.
- Predictive efforts explore what will occur.
- Prescriptive efforts identify what should occur (given the optimization algorithms employed).

One tool providing predictive capabilities was recently unveiled by the giant retailer, Amazon (Bensinger 2014). This patented tool would enable Amazon managers to undertake what it calls "anticipatory shipping," a method to start delivering packages even before customers click "buy." Amazon intends to box and ship products it expects customers in a specific area will want but have not yet ordered. In deciding what to ship, Amazon's analytical process considers previous orders, product searches, wish lists, shopping-cart contents, returns, and even how long an Internet user's cursor hovers over an item.

Relative particularly to agricultural applications and analytics, two key points warrant specific consideration:

- The first continues the veracity discussion introduced in the prior section. Of course, it is prudent to strive to capture and use data which is accurate. However, perfect data generally is expensive to acquire and, given big data approaches, often is not necessary to produce information that can improve decision making.

- For example, it might be technically possible to put sensors to measure actual traffic along every mile of every road in the country. However, the cost of the sensors and the underlying system to aggregate and communicate that information has been prohibitively expensive. However, the use of proxy information (primarily from cell phones and sensors put in place for other purposes) allows for useful predictions of real-time traffic conditions to be created and communicated at low cost.
- Although there are numerous mathematical and statistical approaches available to data scientists, a key principal is the use of Bayesian inference and conditional probabilities (Polson and Scott 2018). Essentially, through the analysis of very large amounts of relevant data, analysists can predict with confidence that the presence of certain factors indicates that the condition of interest exists. As a simple example, if it is 8 am on a weekday morning (that is not a holiday) and cell phone signals along a major highway are moving very slowly from one tower to the next, it is likely that there is heavy traffic along that highway.
- Of course, such a prediction is probabilistic and may not be accurate in each circumstance, especially if there is a change in the underlying conditions. However, with careful analysis and implementation, data that is not perfect can be effectively employed to improve decision making in agriculture. For example, consider the large maize farmer who receives satellite maps of the fields for which the farmer is responsible. Colors are used to identify conditions in the field, with green indicating heavy vegetative growth. As one farmer reported to the chapter's author, an area marked in heavy green means either that the crop is doing really well or that there is a heavy infestation of weeds. In either case, it is worth the farmer's time to physically investigate.
- In agriculture, as in most fields, descriptive efforts have been most common and even those are relatively infrequent. However, within production agriculture, knowing what has occurred – even if very accurately and precisely – may not provide useful insights as to what should be done in the future.
- Production agriculture is complex, where biology, weather, and human actions interact. Science-based methods have been employed to discern why crop and livestock production occurs in the manner in which they do. Indeed, relative to the big data topic, it might be useful to consider this as the "small data" process.
- The process starts with lab research employing the scientific method as a systematic process to gain knowledge through experimentation. Indeed, the scientific method is designed to ensure that the results of an experimental study did not occur just by chance (Herren 2014). However, results left in the lab do not lead to innovation and progress in the farm field. In the United States, the USDA, Land Grant universities, and the private sector have collaborated to exploit scientific advances. A highly effective, but distributed, system emerged where knowledge gained in the laboratory was tested and refined on experimental plots and then extended to agricultural producers.

- In agriculture, therefore, knowledge from science will need to be effectively integrated within efforts to accomplish the goals of predictive and prescriptive analytics. Even with this additional complication, the potential of tools based upon emerging data science capabilities offers significant promise to more effectively optimize operations and create value within the agricultural sector.

8.5 Digital Agriculture and the Food System

To this point, this chapter has focused on individual technologies and concepts that can affect the manner in which big data and digital technologies affect agricultural innovation. This section will attempt to depict a more unified picture of the future setting that we might call digital agriculture. First, the emphasis will continue at the level of production agriculture. The focus will emphasize managerial capabilities, which in most cases will rely upon multiple technical factors. Illustrative examples will be provided. Second, production agriculture is just one component of the broader food system. A system which increasingly is employing digital technologies and big data to improve efficiency and effectiveness. Linkages between those efforts and implementation within production agriculture will be explored in the section's second segment. The role of societal expectations for that system will be discussed as well.

8.5.1 Components of a Potential Digital Agriculture

This section will attempt to paint a picture of the multiple components that could inform farmers in tomorrow's digital agriculture. This is not intended to be a prediction, as these components currently are being employed to some extent. Rather, the discussion will hopefully provide insights as to the potentials that exist as integration across technical capabilities occurs. Of course, one needs to keep in mind the reality that just because something is technically possible, there may not be sufficient justification for managers to adopt that innovation.

Figure 8.3 graphically identifies several components that could form digital agriculture. Examples of each will be provided to illustrate their potential application.

Precision agriculture has been discussed at length previously. While routinely employed by many farmers, advances in technical capabilities are continually offering new opportunities. For example, sensors embedded in the maize planter now can sense soil moisture, depth, and other factors in each furrow to optimize the placement of seed as planting is being done.

Similarly prior discussion addressed big data applications. Farmers now can subscribe to services that provide extremely localized weather information and/or receive agronomic guidance based upon insights gained from analysis of production on thousands of acres, in addition to their own experience. As more data is captured and as research and development continues, the capabilities of such services are likely to increase.

The Internet of Things is a concept closely linked to precision agriculture, although it is worthy of separate consideration. Sensors that monitor conditions in the field to inform irrigation decisions are one example. Similarly sensors in grain bins can continually monitor conditions in the bin. In both instances, the associated systems can inform managers of the actual and past status, can initiate action when warranted, and can record information that can be employed (possibly with big data approaches) to improve the algorithms within the system.

A key difference between factory-based manufacturing and much of agriculture is that agriculture occurs in the open and across significant distances. Therefore monitoring what is happening as it happens has been a historic challenge on the farm. As noted previously remote sensing is being employed to overcome this constraint. The source of the data can be from satellites, fixed wing aircraft, UASs, stationary devices, or some combination of them. For example in Australia, efforts are underway which link satellite monitoring of pasture conditions with sensors that monitor animal weight. Algorithms are being estimated which use that data to recommend when to move animals from a paddock that is in danger of being overgrazed to another more suitable one. Applications in developing countries are particularly exciting as the prior methods of gathering information have been both expensive and insufficient. Possibly, remotely sensed data, in combination with other information sources, can improve agricultural and food systems in those settings. A marked improvement might be possible in a fashion similar to the way cell phones markedly improved communications in developing countries.

This chapter's second section described a story from the 1950s to illustrate that farmers have always wanted to use evidence from their operations to improve the farm's performance. The high cost or infeasibility of measurement historically limited their capabilities to do so. Labelled here as on-farm research, there appears to be is considerable innovation and experimentation focused on the means by which farmers can apply the digital technologies to learn how to improve their own operations (Grains Research Development Council 2016). This interest is being expressed in terms of actions by individual farmers, efforts of groups of farmers (including cooperatives), and in collaboration with input providers. These efforts are exciting because of the possibility of gain, but also because of the potential for enhanced managerial control which previously was never available to farmers.

Finally, coordination is a critically important aspect of digital agriculture. Such coordination may involve relatively simple technologies. For example, a group of woman farmers in an African country who learn that the trader can pay higher prices for their chickens if they use a cell phone to inform the trader when there are enough chickens available to fill the trader's truck. In contrast, coordination may involve combining localized weather forecasts with a logistics model of the most efficient use of equipment for a large farming operation.

8.5.2 *Digital Technologies Throughout the Food System*

Figure 8.4 provides a high-level view of the key subsectors within agriculture that has proved useful for consideration of future competitive dynamics relating to big data. In that graphic, the genetics subsector is separately identified because of its linkages with big data. A number of firms in that category have capabilities to operate as input suppliers as well. The input supply category refers to providers of equipment, seed, fertilizer, and chemicals to farmers as well as providers of financial and managerial services. The production agriculture segment is comprised of farming firms, which can range from low-resourced smallholders to family corporations to subsidiaries of major corporations. The first handler segment refers to firms which aggregate, transport, and initially process agricultural produce but do not directly market to consumers. The final segment relates to food manufacturers and retailers. These types of activities are combined here because of their common interest in employing big data tools to better understand consumers.

From a strategic perspective, it is important to stress that big data tools already are extensively employed, particularly at both "ends" of the sector. Firms at the food manufacturing and the food retailing levels expend considerable resources to continually develop a better understanding of consumers. Insights gained through application of big data analytics can allow managers both to anticipate and respond to consumer concerns. Far upstream in the sector, bioinformatics and other big data tools are employed to accelerate research and development processes, advancing genomic capabilities of the sector. Figure 8.4 identifies, at a general level, key interests that "naturally" reside within each subsector and have the potential to be important within big data applications.

Agricultural operations occur across time and space. Therefore, the logistics of providing inputs, production, and aggregating output consume considerable

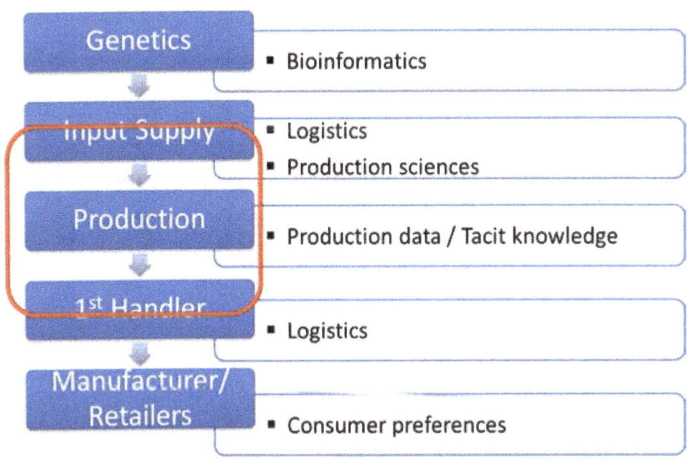

Fig. 8.4 Subsectors and their key strategic interests relating to big data

resources. Advances in information and communication technology combined with big data analytics offer the potential to reduce the amount of resources needed. Deadweight loss is a term that describes system inefficiencies that can be reduced by enhanced coordination within and between firms. Even in advanced agricultural settings, reduction of deadweight loss is perceived to be an attractive potential use of big data innovations.

In this context, deadweight loss refers to the processes by which inputs and outputs are delivered (when and where). A more intriguing issue for many is whether application of big data can fundamentally alter decision making as to "what" should be done. Can we further optimize the biology of agricultural production, especially in the context of the larger food and agricultural system? Earlier it was noted that new sensing technologies offer the potential to monitor and document what actually occurs as agricultural production takes place. The resulting data potentially would be available at never before levels of detail, in terms of time and space, and at low cost. Furthermore, analytic capabilities could combine diverse sources of data to discern previously unknown patterns and provide insights not available previously.

A result of application of these innovations would be optimization of agricultural production systems, simultaneously reducing its environmental impact and improving profitability. There are two interrelated factors that need to be addressed in considering the possible evolution of this optimization:

- Production agriculture involves biologic processes subject to considerable uncertainty. Therefore, even if one knows exactly *what* occurred in one production season and what actions would have optimized performance under those circumstances, that information may not be a good predictor of what actions should occur in the next season.
  - Some assert that capturing massive amounts of agricultural data, spread across large geographic regions, will provide sufficient information so that the effects of weather and location can be estimated. Doing that would enable big data analytics to answer the question, "why does production variability occur?"
  - Agricultural science has been devoted to discerning the why of agricultural production. Rather than solely relying on big data analytics, others assert that agricultural science techniques and knowledge will need to be integrated within big data techniques to truly optimize system performance.
- In most systems of agricultural production today, even the knowledge of what occurred does not necessarily reside within one organization. Furthermore, as was noted for precision agriculture, individual entities at the production level typically do not have the scale to produce sufficient data nor to have the capabilities needed to analyze that data.

Because of these two factors, collaboration across organizational boundaries will be required to fully exploit the potential benefits of big data's application to agriculture. A host of factors, beyond technological effectiveness, will influence the speed and extent of this exploitation. These relate to intellectual property and competitive dynamics as well as the magnitude of economic benefits available. Such impediments are not insurmountable and can be viewed as much as opportunities as they are impediments. How they are resolved, however, will have a major impact on big data's eventual contribution to performance within agriculture.

Beyond its direct economic impact, society has intense interest in the social and environmental effects of agriculture. Food safety and security are of public interest in every society. Interest in mitigating negative environmental impacts of agricultural operations is increasingly a concern and that concern is not constrained to just citizens in developed nations. In addition to public sector interest, some consumer segments express interest and concern regarding the practices and methods employed to produce food. Therefore, in addition to public sector-based regulation, documentation as to practices employed is increasingly being required by private sector food manufacturers and retailers.

Interestingly, technological innovations, such as those noted previously, have the potential to provide much better evidence as to their societal and environmental effects. These include both tools to more precisely measure and monitor and analytical methods to better understand and predict effects.

8.6 Final Remarks

Historically, agricultural innovation has been a key factor in the improvement of individual and societal well-being. While much progress has been achieved through that innovation, more needs to be accomplished especially as the world addresses the needs of a growing population while simultaneously reducing the stress that agricultural production can have on the natural environment. This chapter explores factors affecting the extent to which effective implementation of digital technologies and big data can contribute to that urgently needed agricultural innovation.

References

Barth D (2009) The bright side of sitting in traffic: crowdsourcing road congestion data. Retrieved February, 2, 2016, from Google's Official Blog Web site: https://googleblog.blogspot.com/2009/08/bright-side-of-sitting-in-traffic.html

Bensinger G (2014) Amazon wants to ship your package before you buy it. The Wall Street Journal. Retrieved February, 2, 2016, from the Wall Street Journal Web site: http://blogs.wsj.com/digits/2014/01/17/amazon-wants-to-ship-your-package-before-you-buy-it/

Brynjolfsson E, McAfee A (2012) Big data: the management revolution. Harv Bus Rev 90(10):60–68

Chen M, Mao S, Liu Y (2014) Big data: a survey. Mob Netw Appl 19(2):171–209

Coble K, Ahearn M, Sonka S, Griffin T, Ferrell S, McFadden J, Fulton J (2016) Advancing U.S. agricultural competitiveness with big data and agricultural economic market information, analysis, and research. The Council on Food, Agricultural and Resource Economics, Washington, D.C. 17 pp

Cooper M, Mell P (2012) Tackling big data [PDF document]. Retrieved February 2, 2016, from National Institute of Standards and Technology Web site: http://csrc.nist.gov/groups/SMA/forum/documents/june2012presentations/fcsm_june2012_cooper_mell.pdf

Davenport TH (2013) Analytics 3.0. Harv Bus Rev 91(12):64–72

Dhar V (2013) Data science and prediction. Commun ACM 56(12):64–73

Erickson B, Lowenberg-DeBoer J, Bradford J (2017) 2017 precision agricultural services dealership survey. Purdue University, West Lafayette, 27 pp

Gandomi A, Haider M (2015) Beyond the hype: big data concepts, methods, and analytics. Int J Inf Manag 35:137–144

Gartner IT Glossary (2012) Big data. Retrieved June, 5 2015, from Gartner Web site: http://www.gartner.com/it-glossary/big-data

Grains Research Development Corporation (2016) Calculating return on investment for on farm trials. Canberra. 20 pp.

Hardy Q (2014) G.E. opens its Big Data platform. Retrieved February 2, 2016, from The New York Times Web site: http://bits.blogs.nytimes.com/2014/10/09/ge-opens-its-big-data-platform/?_r=2

Herren RV (2014) The science of agriculture a biological approach, 4th edn. Cengage Learning, Independence

IBM (2012) What is big data? Retrieved November 22, 2015, from IBM Web site: https://www-01.ibm.com/software/data/bigdata/what-is-big-data.html

King R (2013) Farmers experiment with drones. Retrieved February 2, 2016, from The Wall Street Journal Web site: http://blogs.wsj.com/cio/2013/04/18/farmers-experiment-with-drones/

Manyika J, Chui M, Brown B, Bughin J, Dobbs R, Roxburgh C, Byers A (2011) Big Data: the next frontier for innovation, competition, and productivity. Retrieved February 2, 2016 from McKinsey & Company Web site: http://www.mckinsey.com/insights/business_technology/big_data_the_next_frontier_for_innovation

Marr B (2014) How GE is using Big Data to drive business performance. Retrieved February 2, 2016, from SmartDataCollective Web site: http://www.smartdatacollective.com/bernardmarr/229151/how-ge-using-big-data-drive-business-performance

Meadows DH, Meadows DL, Randers J, Behrens WW III (1972) The limits to growth. Universe Books, New York

National Research Council (1997) Precision agriculture in the 21st century: geospatial and information technologies in crop management. National Academies Press, Washington, D.C.. Accessed October 30, 2015. http://www.nap.edu/catalog/5491/precision-agriculture-in-the-21st-century-geospatial-and-information-technologies

National Science Foundation (2012) Core techniques and technologies for advancing big data science & engineering. Retrieved February 2, 2016, from the National Academies Press Web site: http://www.nsf.gov/pubs/2012/nsf12499/nsf12499.htm

Polson N, Scott J (2018) AIQ. St. Martin's Press, New York. 262 pp

Remote sensing-based information and insurance for crops in emerging economics [RIICE] (2013) About RIICE. Retrieved February 2, 2016, from International Rice Research Institute Web site: http://www.riice.org/about-riice/

Sonka ST (2015) Big Data: from hype to agricultural tool. Farm Policy J 12:1–9

Sonka ST (2016) Big data: fueling the next evolution of agricultural innovation. J Innov Manag 4:114–136

Sonka ST, Coaldrake KF (1996) Cyberfarm: what does it look like? What does it mean? Am J Agric Econ 78(5):1263–1268

Watson HJ (2013) The business case for analytics. BizEd 12(3):49–54

Weersink A, Fraser E, Pannell D, Duncan E, Rotz S (2018) Opportunities and challenges for big data and agricultural and environmental analysis. Ann Rev Resour Econ 10:19–37

Wiseman L, Sanderson J, Robb L (2018) Rethinking Ag data ownership. Farm Policy J 15:71–77

Permissions

All chapters in this book were first published by Springer; hereby published with permission under the Creative Commons Attribution License or equivalent. Every chapter published in this book has been scrutinized by our experts. Their significance has been extensively debated. The topics covered herein carry significant information for a comprehensive understanding. They may even be implemented as practical applications or may be referred to as a beginning point for further studies.

The contributors of this book come from diverse backgrounds, making this book a truly international effort. We would like to thank all the contributing authors for lending their expertise to make the book truly unique. They have played a crucial role in the development of this book. Without their invaluable contributions this book wouldn't have been possible. They have made vital efforts to compile up to date information on the varied aspects of this subject to make this book a valuable addition to the collection of many professionals and students.

This book was conceptualized with the vision of imparting up-to-date and integrated information in this field. To ensure the same, a matchless editorial board was set up. Every individual on the board went through rigorous rounds of assessment to prove their worth. After which they invested a large part of their time researching and compiling the most relevant data for our readers.

The editorial board has been involved in producing this book since its inception. They have spent rigorous hours researching and exploring the diverse topics which have resulted in the successful publishing of this book. They have passed on their knowledge of decades through this book. To expedite this challenging task, the publisher supported the team at every step. A small team of assistant editors was also appointed to further simplify the editing procedure and attain best results for the readers.

Apart from the editorial board, the designing team has also invested a significant amount of their time in understanding the subject and creating the most relevant covers. They scrutinized every image to scout for the most suitable representation of the subject and create an appropriate cover for the book.

The publishing team has been an ardent support to the editorial, designing and production team. Their endless efforts to recruit the best for this project, has resulted in the accomplishment of this book. They are a veteran in the field of academics and their pool of knowledge is as vast as their experience in printing. Their expertise and guidance has proved useful at every step. Their uncompromising quality standards have made this book an exceptional effort. Their encouragement from time to time has been an inspiration for everyone.

The publisher and the editorial board hope that this book will prove to be a valuable piece of knowledge for students, practitioners and scholars across the globe.

Index

www.ingramcontent.com/pod-product-compliance
Lightning Source LLC
LaVergne TN
LVHW051045220926
848485LV00001BA/16

* 9 7 8 1 6 3 9 8 9 7 2 9 2 *